Practical Astronomy

Springer
London
Berlin
Heidelberg
New York
Barcelona
Budapest
Hong Kong
Milan
Paris
Santa Clara
Singapore
Tokyo

Other titles in this series

The Observational Amateur Astronomer
Patrick Moore (Ed.)

The Modern Amateur Astronomer
Patrick Moore (Ed.)

Telescopes and Techniques
Chris Kitchin

Small Astronomical Observatories
Patrick Moore (Ed.)

The Art and Science of CCD Astronomy
David Ratledge (Ed.)

The Observer's Year
Patrick Moore

Seeing Stars
Chris Kitchin and Robert W. Forrest

Photo-guide to the Constellations
Chris Kitchin

The Sun in Eclipse
Michael Maunder and Patrick Moore

Amateur Telescope Making
Stephen F. Tonkin

Observing Meteors, Comets, Supernovae and other Transient Phenomena
Neil Bone

Astronomical Equipment for Amateurs
Martin Mobberley

Software and Data for Practical Astronomers

The Best of the Internet

David Ratledge

Springer

David Ratledge, BSc, CEng, MICE
24 The Common, Adington, Chorley,
Lancashire PR7 4DR, UK

Cover illustration: Background image: Skymap by Chris Marriott; Left inset: Lunar Occultation Workbench by Eric Limburg; Right inset: NEWT for Windows by Dale Keler

ISBN 1-85233-055-4 Springer-Verlag London Berlin Heidelberg

British Library Cataloguing in Publication Data
Ratledge, David, 1945-
Software and data for practical astronomers : the best of the Internet. - (Practical astronomy)
1.Internet (Computer network) 2.Astronomy - Data processing
3.Astronomy - Computer network resources
I.Title
004.6'78'024'521
ISBN 1852330554

Library of Congress Cataloging-in-Publication Data
Ratledge, David, 1945-
Software and data for practical astronomers : the best of the Internet / David Ratledge.
p. cm. -
(Practical astronomy)
ISBN 1-85233-055-4 (alk. paper)
1. Astronomy-Computer network resources. 2. Internet (Computer network) I. Title. II. Series.
QB14.3.R37 1998 98-29961
025.06'52-dc21 CIP

Printed in Great Britain

Typeset by EXPO Holdings, Malaysia
Printed at the Cromwell Press, Wiltshire
58/3830-543210 Printed on acid-free paper

Acknowledgements

A book like this could not have been written if many people had not put a considerable amount of effort into publishing and distributing astronomical information on the Internet. It is probably the case now that there are more private individuals doing so than professional bodies. However, whether it be amateur or professionally produced, we live in an age where information has never been more plentiful and available. I shall try to thank everyone who either contributed data and/or software, or gave their permission for me to include information here. Because of the sheer numbers involved, I will probably have omitted one or two. This is partially due to my memory but also to the fact that many hide anonymously under the title "webmaster". Whichever the reason, my apologies to anyone omitted.

So here goes – without the help of and/or contributions from the following, this book could not have happened and my thanks to them all:

Irene Szewczuk, Patrick Wallace, Chris Marriott, Nick Thompson, Stephen Schimpf, Jeff Bondono, Bob Erdmann, Steve Coe, Jon Giorgini, Alan Chamberlin, Gareth Williams, Monzur Ahmed, Tim Puckett, Fred Espenak, Ray Robinson, Eric Limburg, Michael Perryman, Herbert Raab, Bev Ewen-Smith, Jerry Gunn, Ingrid Siegert, Barry Lasker, Kirk Goodall, Chris d'Aquin, Richard Berry, Chritian Buil, Christer Strandh, Steve Tonkin, Tim Poulsen, Jim Burrows, Jean Prideaux, Donald Wright, Jerry Wright, Webster Cash, Larry Phillips, Gerald Pearson, Dale Keler, Dick Suiter, Mel Bartels, Bruce Sayre, Gerald Bramall, Brian Webber, Dave Harvey, Paul Traufler, Dave Cappellucci, Bradford Beyr, Richard Dreiser, Sally Houghton, Richard West, Ray Riggs, Brian Colville, Omar Thameen, Wendy & Jess Diard, David Irizarry, and James Webb.

On the production front, my thanks are due to John and Jon. John is John Watson of Springer who has

always been around to help when things got tricky. Jon is Jon Ritchie who did the big downloads and saved me a fortune on my phone bill. Last but not least, my thanks are due to my wife for patiently proof reading the manuscript. Needless to say any remaining errors are probably ones I put in afterwards and are not her responsibility.

David Ratledge
24 The Common,
Adlington,
Chorley,
Lancs. PR7 4DR
England

Email: david_ratledge@compuserve.com

Contents

Chapter 1 Introduction

Background

The Internet could have been invented for Astronomy. No longer the preserve of research institutions and educational establishments, this all pervading computer network now links astronomers, both professional and amateur, all over the world and connects them to the data and information they need. Email provides the former and the World Wide Web (WWW) the latter. I can recall the voyager missions and all those fabulous images. If only we could have got our hands on them, rather than waiting for the press to publish a few extracts. That has all changed now. The minute that new images or data is released, it will probably be published on the Internet, making it simultaneously available all over the world. That is to those with Internet access. And in its entire original form, not edited or abridged by a journalist who doesn't understand astronomy.

The WWW, although a relatively recent innovation on the Internet, now dominates it as a method of distributing data. Its combination of text, hyperlinks and graphics has become the standard method of publishing. It is not difficult to see why. The ease of use and "friendliness" of browsers such as Netscape Navigator and Internet Explorer means that even novices can quickly get up to speed and "surf the net" for the information they want. No need for reading computer manuals, everything is intuitive and straightforward. Literally "point and click".

It doesn't matter what astronomical subject you are interested in. Even for the most obscure of subjects there is certain to be a site, probably several, covering it in complete detail. Telescope making? – no problem you will have ideas and inspiration aplenty. Need an image of galaxy or quasar? – a few clicks and you will have one. The list is endless. To reach a site all the surfer has to do is type in the WWW address, known as a URL (uniform resource locator) and the site will magically appear, no matter where it is located in the World. And all for the price of a local telephone call.

However, the WWW has become a victim of its own success. As more and more people, not just us astronomers, discover the Internet then popular sites get swamped and struggle to serve their audience. Downloads become painstakingly slow and, in the UK with our telephone charges, extremely expensive. Downloading just one NASA video could cost several £s. Another problem is sorting the good from the bad. There is no veto on publishing and quality varies enormously. Now the good news, with the help of this book you will hopefully have short cuts to some of the best sites plus, on the CD-ROM accompanying it, some of the most useful data and software already downloaded.

Figure 1.1. The Digital AltaVista Search Site with results from query "Astronomical Observatories".

Finding Your Way Around

We learnt earlier that by simply typing in the URL of a site we would be immediately transported to it. But suppose we don't know the site's URL or even if one exists. All is not lost however, as several philanthropic companies supply search facilities (known as Search Engines) for the whole Internet. They are not philanthropic really, advertising usually funds their operations but the result is that we get a wonderful free service from them.

Operating them could not be simpler. On their sites the surfer is presented with a form where keywords are simply entered and, when the "search" button is clicked, the entire Internet is trawled for the best matches. They do not actually trawl on the fly – that would be far too slow. They have already indexed all WWW pages and it is their indexes they search. This pre-indexed search is very quick but it does mean that there is a significant delay between a new page being published and the search engine knowing about it. The results of the search are presented, usually in batches

Figure 1.2. The Yahoo Astronomy Index.

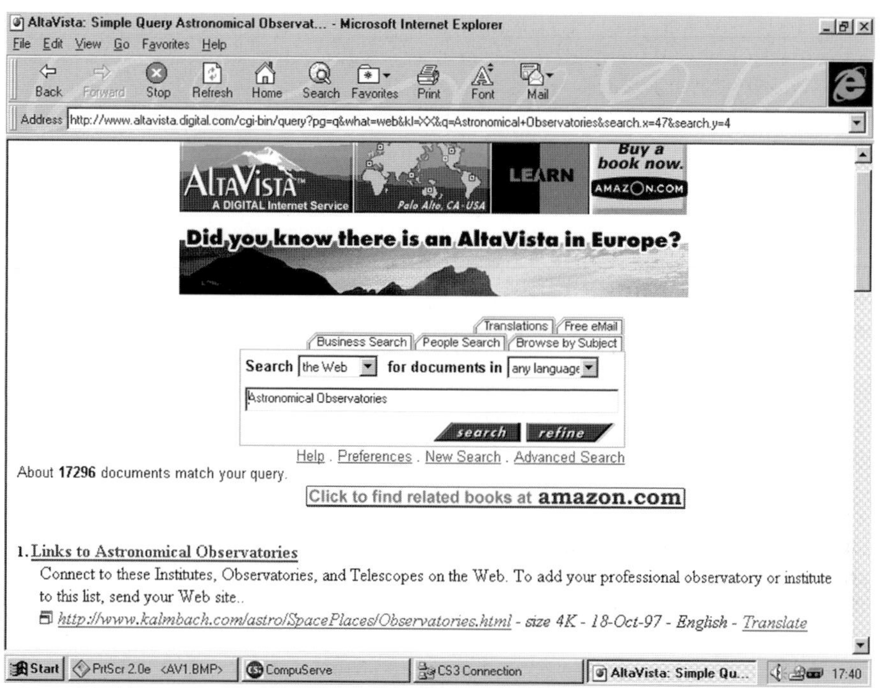
AltaVista: Simple Query Astronomical Observat... - Microsoft Internet Explorer
File Edit View Go Favorites Help
Back Forward Stop Refresh Home Search Favorites Print Font Mail
Address http://www.altavista.digital.com/cgi-bin/query?pg=q&what=web&kl=XX&q=Astronomical+Observatories&search.x=47&search.y=4
AltaVista A DIGITAL Internet Service
Palo Alto, CA - USA
LEARN
Buy a book now. AMAZON.COM
Did you know there is an AltaVista in Europe?
Translations Free eMail
Business Search People Search Browse by Subject
Search the Web for documents in any language
Astronomical Observatories
search refine
Help . Preferences . New Search . Advanced Search
About 17296 documents match your query.
Click to find related books at amazon.com
1. Links to Astronomical Observatories
Connect to these Institutes, Observatories, and Telescopes on the Web. To add your professional observatory or institute to this list, send your Web site..
http://www.kalmbach.com/astro/SpacePlaces/Observatories.html - size 4K - 18-Oct-97 - English - Translate
Start PrtScr 2.0e <AV1.BMP> CompuServe CS3 Connection AltaVista: Simple Qu... 17:40

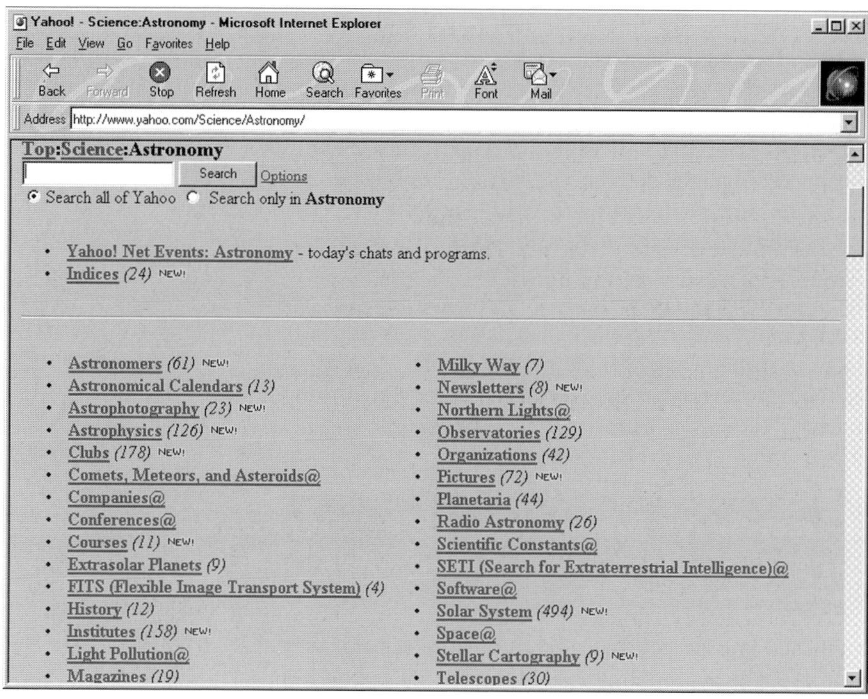
Yahoo! - Science:Astronomy - Microsoft Internet Explorer
File Edit View Go Favorites Help
Back Forward Stop Refresh Home Search Favorites Print Font Mail
Address http://www.yahoo.com/Science/Astronomy/
Top:Science:Astronomy
Search Options
Search all of Yahoo Search only in Astronomy
Yahoo! Net Events: Astronomy - today's chats and programs.
Indices (24) NEW!
Astronomers (61) NEW!
Astronomical Calendars (13)
Astrophotography (23) NEW!
Astrophysics (126) NEW!
Clubs (178) NEW!
Comets, Meteors, and Asteroids@
Companies@
Conferences@
Courses (11) NEW!
Extrasolar Planets (9)
FITS (Flexible Image Transport System) (4)
History (12)
Institutes (158) NEW!
Light Pollution@
Magazines (19)
Milky Way (7)
Newsletters (8) NEW!
Northern Lights@
Observatories (129)
Organizations (42)
Pictures (72) NEW!
Planetaria (44)
Radio Astronomy (26)
Scientific Constants@
SETI (Search for Extraterrestrial Intelligence)@
Software@
Solar System (494) NEW!
Space@
Stellar Cartography (9) NEW!
Telescopes (30)

Figure 1.3. The Lycos Search Site – results of search on "Astronomical Observatories".

of ten, with the best match first. When framing your query be as specific as possible. Too vague a query and tens of thousands of "matches" will be returned.

It is often a good idea to use more than one search engine as they do return different results, or at least different priority orders. The companies don't reveal how they prioritise – if they did web authors would design their pages so that the search engines would find them first, so defeating the whole object!

My personal preference is for AltaVista first, followed by Lycos and Yahoo. AltaVista runs on some of the most powerful processors in the world and it is relatively quick even at busy times. It can even translate pages in some foreign languages to English – amazing! Yahoo is noteworthy for having a specific Astronomy index.

General Sites with Good Links

A supplement to the search engine is to "bookmark" useful sites with good links to both new and established

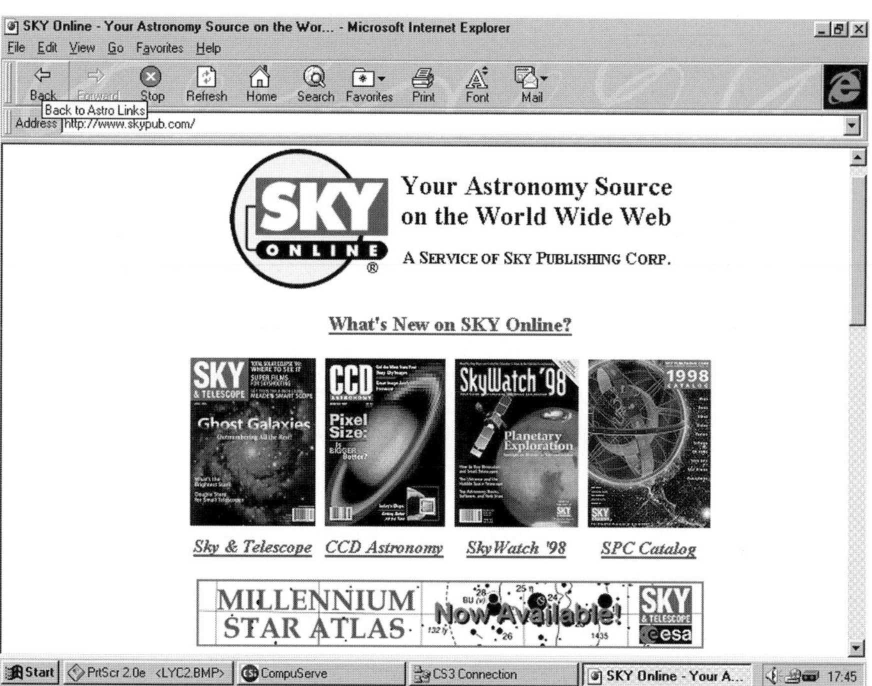

Figure 1.4. Sky Publishing's Homepage – a good place to start.

sites. Both Netscape Navigator and Internet Explorer have facilities for saving your favourite web sites. There is then no need to type anything in at all. A mouse click recalls the address. In addition, you can make your number one favourite site your default start page. Then every time the browser is loaded it will automatically go to this page. My personal default is the Sky Publishing Corporation's Homepage, from the publishers of *Sky & Telescope* magazine. They have an excellent links section and if a new site, such as a space mission, appears then they will have a link to it. It also publishes a weekly news bulletin which makes regular visiting worthwhile.

For a more British perspective on all astronomical matters then Astronomy Now is also worth bookmarking. Perhaps not the same depth as Sky Publishing but interesting nonetheless.

Starlink

If you would like to know how UK professional astronomers use and share software and data the Starlink Homepage is worth visiting.

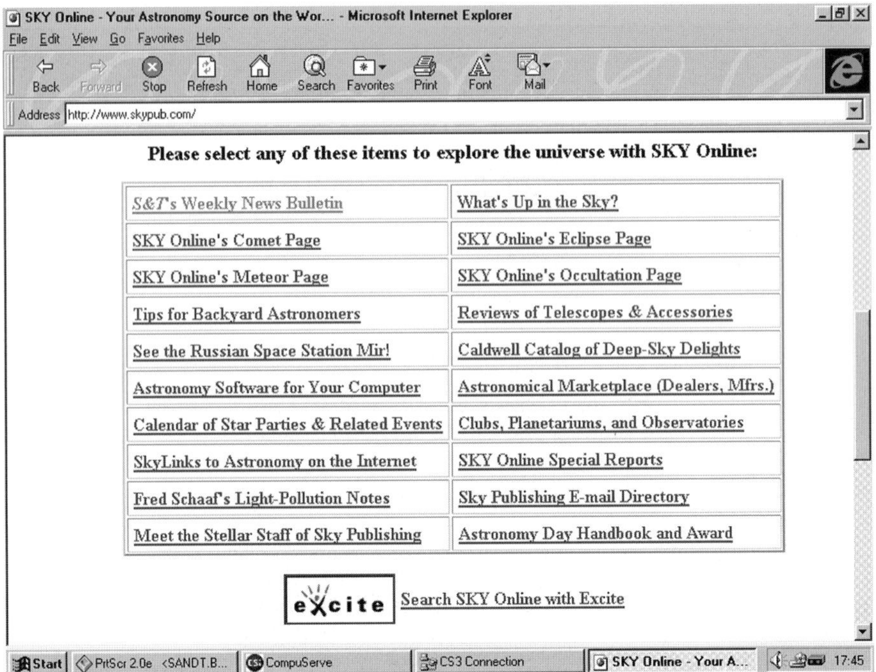

Figure 1.5. Sky Publishing's excellent Links Page.

Starlink's aim is to provide interactive data processing facilities for UK astronomers. This is a shared project with most of the work taking place on university sites. It provides many services including:–

- A comprehensive software collection with over 120 software items
- A distribution service for commercial software
- Hardware, programmers, documentation and technical support.

Although clearly targeted at professional astronomers, Starlink software is available for the use of individual amateur astronomers. However, the software is Unix based so if you plan to use it on a PC then you are recommended to use the Linux operating system. If you would like to know more about Starlink, then their Frequently Asked Questions (FAQ), which runs to 18 pages, is included on the accompanying CD-ROM.

Downloading

Having found what we are looking for, the natural next step is downloading it. This, the most popular of Internet

Figure 1.6. Astronomy Now's Homepage – a British perspective.

activities, was traditionally handled by file transfer protocol (FTP) with its own utilities. However, this is now more commonly, at least for us amateurs, handled by normal web browsers. Unfortunately the most interesting files are often the biggest and download times can be frightening – time is literally money. If using a modern browser to download, an estimation of how long the operation will take will be given (Figure 1.8, *overleaf*). Unfortunately this is only given after the download has begun and the browser has had time to extrapolate an estimated duration time. Give it a few seconds to settle down as initial estimates are usually erratic.

Download times are notoriously difficult to predict. In theory, knowing the file size (not always quoted unfortunately) and our modem connect speed, it should be an easy calculation but life is not that simple. Factors such as how fast your Internet supplier connects you to the Internet, how heavily loaded is the server you are accessing and even the time of day, all have a major bearing on the answer. The downloads for the accompanying CD were done through a high speed academic link called Super Janet. For mere mortals, when you find something large worth downloading, my own recommendation would be to set it going at 7:00 a.m., whilst the Americans are fast asleep!

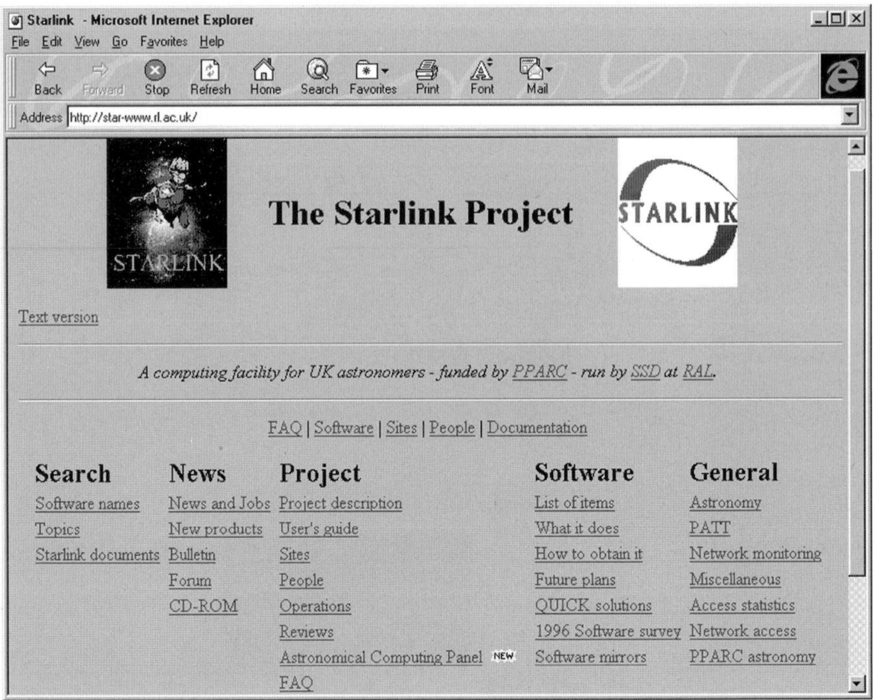

Figure 1.7. The Starlink Homepage.

To speed downloads, many files are compressed. In the PC world these can usually be recognised by the suffix ".zip". However not all Internet files are in PC format – in fact, in the professional world, Unix is still more common as we learnt at Starlink. There are various utilities available to un-compress files such as Pkunzip for DOS, Stuffit Expander and WinZip. A selection is included on the accompanying CD-ROM and you can choose your own favourite..

Finally, new sites are appearing at an ever increasing pace and the best way to keep in touch is by reading

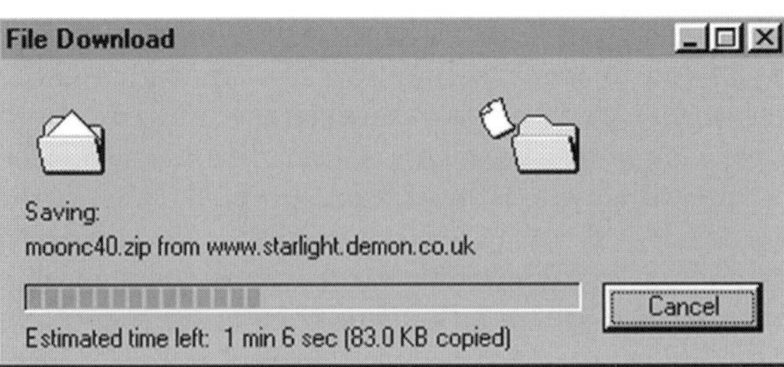

Figure 1.8. Typical download information but the time left only appears after the download has started!

the monthly online articles in most astronomy magazines. *Sky & Telescope* has an excellent comprehensive feature edited by Stuart J. Goldman and I thoroughly recommend it. Different themes are selected each month. The UK magazine *Astronomy Now* also has a regular column by David Johnson which provides a short review of web sites.

URLs Featured in this Chapter

AltaVista Search Engine:
http://www.digital.altavista.com

Yahoo Search Engine:
http://www.yahoo.com

Lycos Search Engine:
http://www.lycos.com

Sky Publishing Corporation:
http://www.skypub.com

Astronomy Now:
http://www.astronomynow.com

Starlink
http://star-www.rl.ac.uk

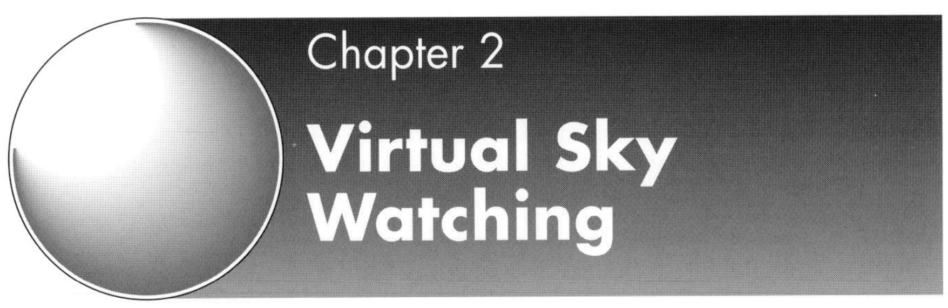

Chapter 2
Virtual Sky Watching

Background

As soon as personal computers were invented it was only natural that the digital equivalent of the planisphere would be created. From those early days of computing, sky simulations have been probably the biggest selling of all astronomical software. No astronomer worth his salt would be without one. All these programs will chart the sky for any date and place on Earth, some even as seen from other planets, although just how useful that would be is a moot point. The number of stars depends on the database included which can vary with versions. The most basic will have the Bright Star Catalogue which has over 9,000 stars down to magnitude 6.5. Next level up is the Smithsonian Astrophysical Observatory (SAO) Catalogue with 250,000 stars to around magnitude 8. Some programs include the Hubble Guide Star Catalogue with 15 million stars to magnitude 15 – enough for most people!

The Moon, planets and comets will be accurately plotted when visible. They identify constellations, stars and thousands of deep sky objects including the exciting Messier ones. Finder charts can be printed out to a quality rivalling the best star atlas. Levels of sophistication vary but even the most humble DOS program will have most of the above features. The advanced ones add more bells and whistles and have features such as accurately predicting and simulating eclipses and holding images which will be displayed when an object is queried.

However, unlike those early days of personal computing when choice was very restricted, today the amateur astronomer suffers from an embarrassment of riches. The choice is bewildering. If only we could try before we buy. With the Internet that becomes a reality! There is much software out there to try for free. They split into broadly three categories, *freeware*, *shareware* and *demos*. Freeware is just that, you are free to use it, usually without reservation. Shareware however is definitely not free although it can appear so. Your use is restricted to a trial period, usually 30 days, after which you must either delete it from your PC or register it and pay the required fee. Demo software is generally a restricted version of the full programme – something is invariably missing e.g. printing may be prohibited. It does however give a feel of how the program will perform on your computer.

A good place to begin your search for the ideal package is on my favourite homepage, Sky Publishing. Here, John Mosley keeps an up to date listing of all kinds of astronomical software in his excellent and comprehensive revue.

Not all the sites listed offer shareware or demos but it is an excellent place to start. If it is only shareware

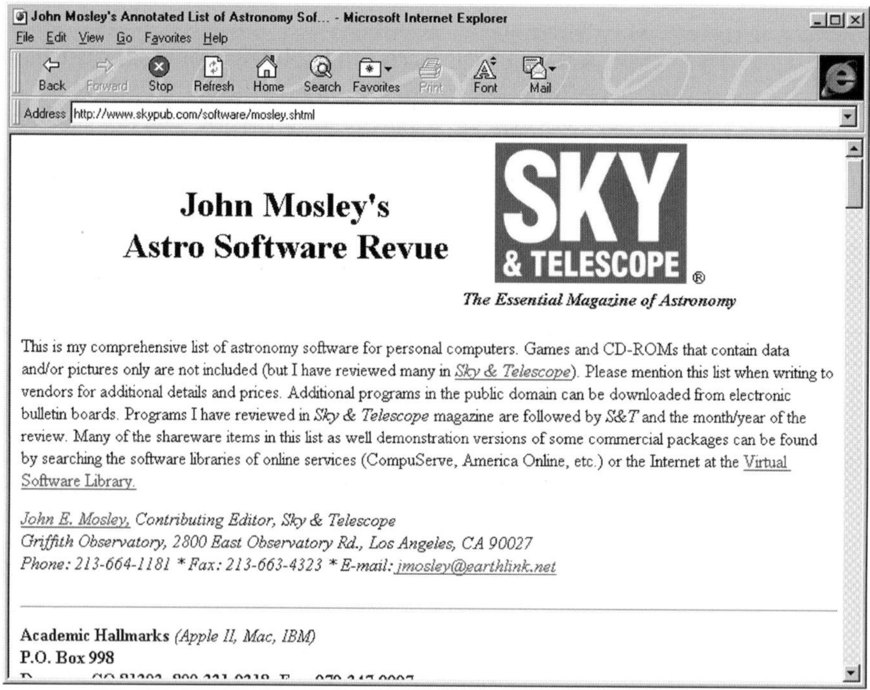

John Mosley's Annotated List of Astronomy Sof... - Microsoft Internet Explorer
File Edit View Go Favorites Help
Back Forward Stop Refresh Home Search Favorites Print Font Mail
Address http://www.skypub.com/software/mosley.shtml
John Mosley's
Astro Software Revue
SKY & TELESCOPE ®
The Essential Magazine of Astronomy
This is my comprehensive list of astronomy software for personal computers. Games and CD-ROMs that contain data and/or pictures only are not included (but I have reviewed many in Sky & Telescope). Please mention this list when writing to vendors for additional details and prices. Additional programs in the public domain can be downloaded from electronic bulletin boards. Programs I have reviewed in Sky & Telescope magazine are followed by S&T and the month/year of the review. Many of the shareware items in this list as well demonstration versions of some commercial packages can be found by searching the software libraries of online services (CompuServe, America Online, etc.) or the Internet at the Virtual Software Library.
John E. Mosley, Contributing Editor, Sky & Telescope
Griffith Observatory, 2800 East Observatory Rd., Los Angeles, CA 90027
Phone: 213-664-1181 * Fax: 213-663-4323 * E-mail: jmosley@earthlink.net
Academic Hallmarks (Apple II, Mac, IBM)
P.O. Box 998

Figure 2.1. John Mosley's comprehensive and up to date review of Astronomical Software.

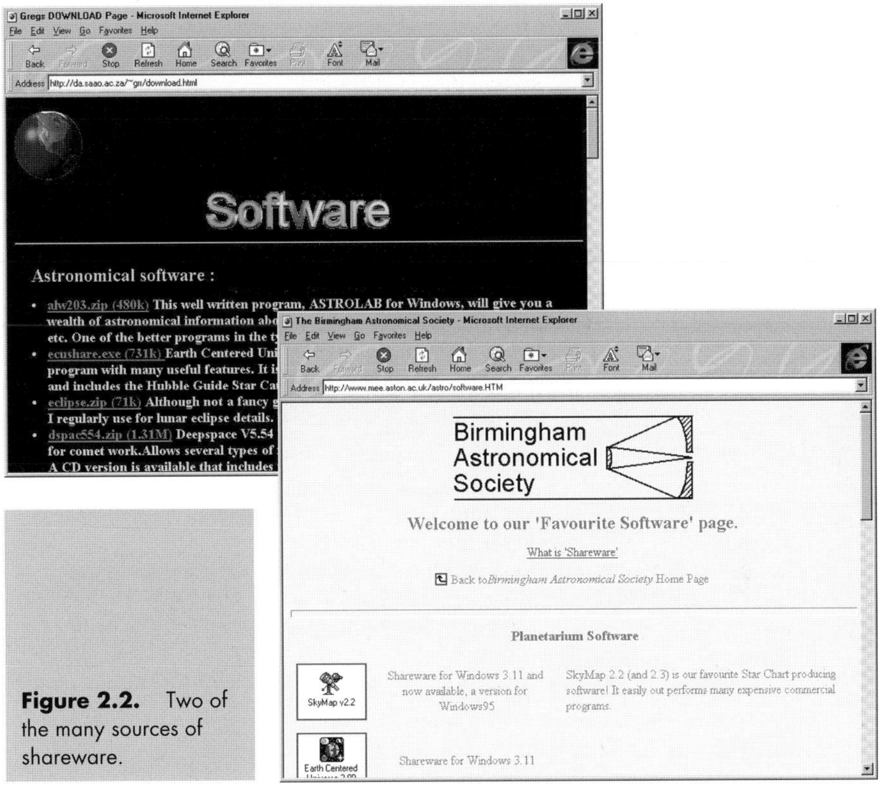

Figure 2.2. Two of the many sources of shareware.

you are interested in (and who isn't) then various official and unofficial sites offer copies of their particular favourites. Two typical sites with software to download are illustrated in Figure 2.2. To find more sites just use the search engines referred to in Chapter 1, using a query such as "Astronomical Shareware" or "Astronomical Freeware" if funds are severely limited. Either way you will be given plenty of choice.

I will now describe in more detail some of the planetarium programs that are included on the CD-ROM accompanying this book. They are all different with their own specialities.

SkyMap

SkyMap is an excellent example of the sky simulation/planetarium/mapping program and what is more, it is written by a British amateur astronomer Chris Marriott. Its forte is producing finder charts, hardly

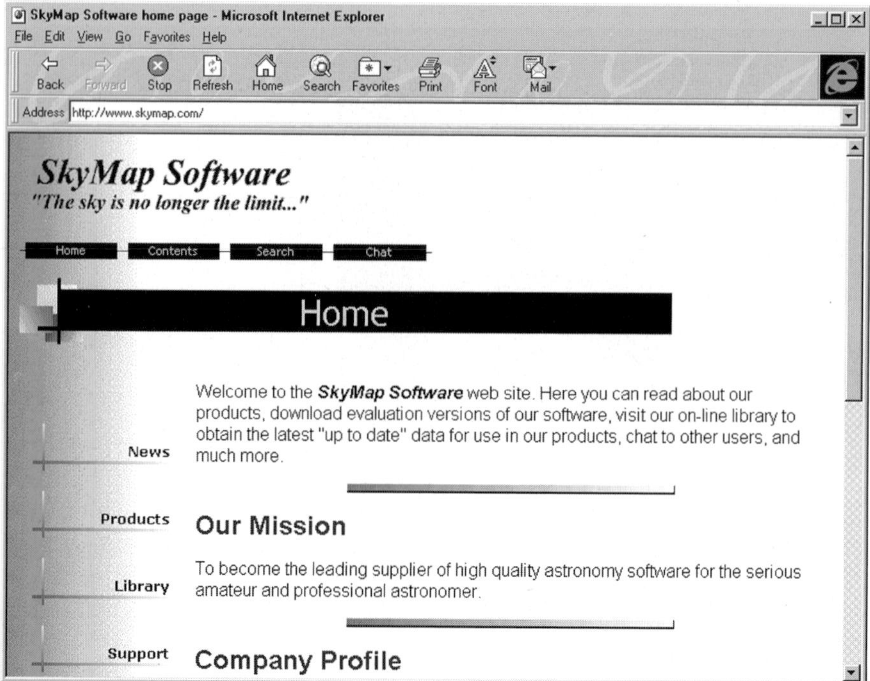
SkyMap Software home page - Microsoft Internet Explorer
File Edit View Go Favorites Help
Back Forward Stop Refresh Home Search Favorites Print Font Mail
Address http://www.skymap.com/
SkyMap Software
"The sky is no longer the limit..."
Home Contents Search Chat
Home
Welcome to the SkyMap Software web site. Here you can read about our products, download evaluation versions of our software, visit our on-line library to obtain the latest "up to date" data for use in our products, chat to other users, and much more.
News
Products
Our Mission
To become the leading supplier of high quality astronomy software for the serious amateur and professional astronomer.
Library
Support
Company Profile

Figure 2.3. The SkyMap Homepage – where you can download the latest comet database.

surprising in view of its name, but it does much more as well. SkyMap is currently available in three versions. Version 2 is for Windows (3.1 and above) and version 3 for Windows95/98. Version 4, known as SkyMap Pro, includes the Hubble Guide Star Catalogue with stars down to 15th magnitude. Version 2 is normally shareware but as a special concession to the Practical Astronomy series, Chris has generously agreed to provide us with a fully registered copy of it (Version 2.2). Just about worth the price of this book alone! You will find it together with the shareware version of 3.1 on the CD-ROM.

The version of SkyMap illustrated here (see Figures 2.4, *opposite*, and 2.5, *overleaf*) is in fact version 2.2, and it is suitable for Windows 3.1 and above. It opens with what is known as an "horizon map" i.e. what the sky would look like when viewed south at the current time, based on a location in the north west of England (where the author of SkyMap happens to live!). However it is an easy procedure to change the location to wherever on Earth you live and this default can be saved. From the "horizon map" it is a simple matter to zoom into any area of interest.

Where SkyMap excels is in detailed finder charts, known as "area maps". The amount of detail, such as

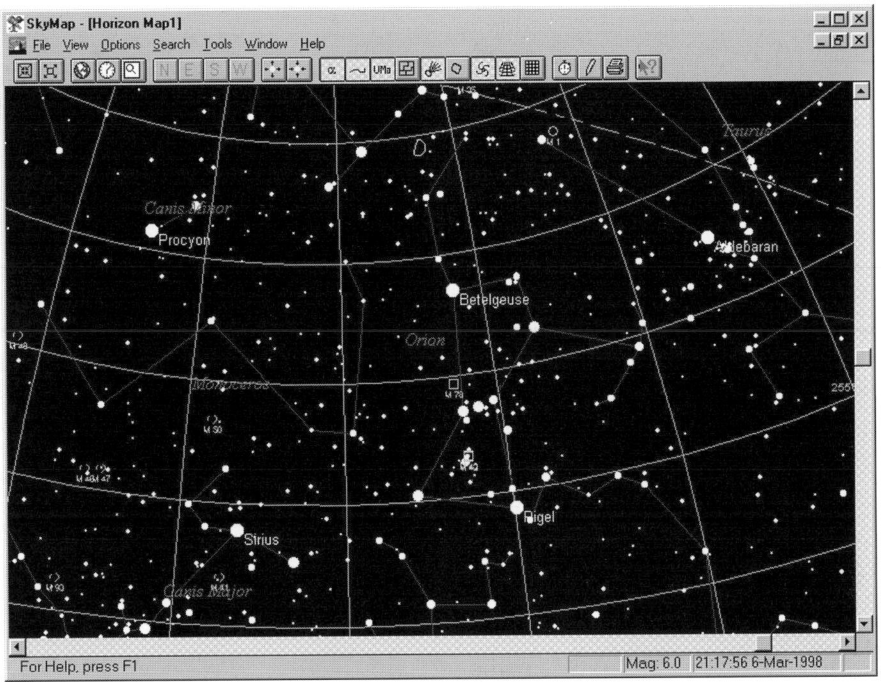

Figure 2.4. SkyMap's opening "horizon" map.

the faintest star shown and the magnitudes of deep sky objects plotted, can be varied to get it just how you want it. Charts can be centred on any object unlike printed star atlases where the object of interest usually falls on the edge of two maps or even worse on the corner of four! For objects moving relative to the stars, such as the planets, asteroids and comets, there is a "track mode" where the path of the object will be plotted over several days. In Figure 2.5 (*overleaf*) the path of Comet Tempel-Tuttle is shown passing Saturn.

For new comets not included in the program's standard database, their orbital elements (see Chapter 4 for where to find them on the Internet) can be entered. Alternatively the SkyMap Homepage keeps an up to date comet database that is small enough to be quickly downloaded. When this is loaded into the SkyMap directory, all new comets are available for accurate plotting. An excellent free service.

For deep sky objects there is a simple "find" command, which works on either the objects catalogue number (Messier or NGC) or its popular name e.g. The Lagoon Nebula. Galaxies are plotted to their catalogue size and orientation. Very useful for CCD imagers who have to orientate their camera to get the best coverage.

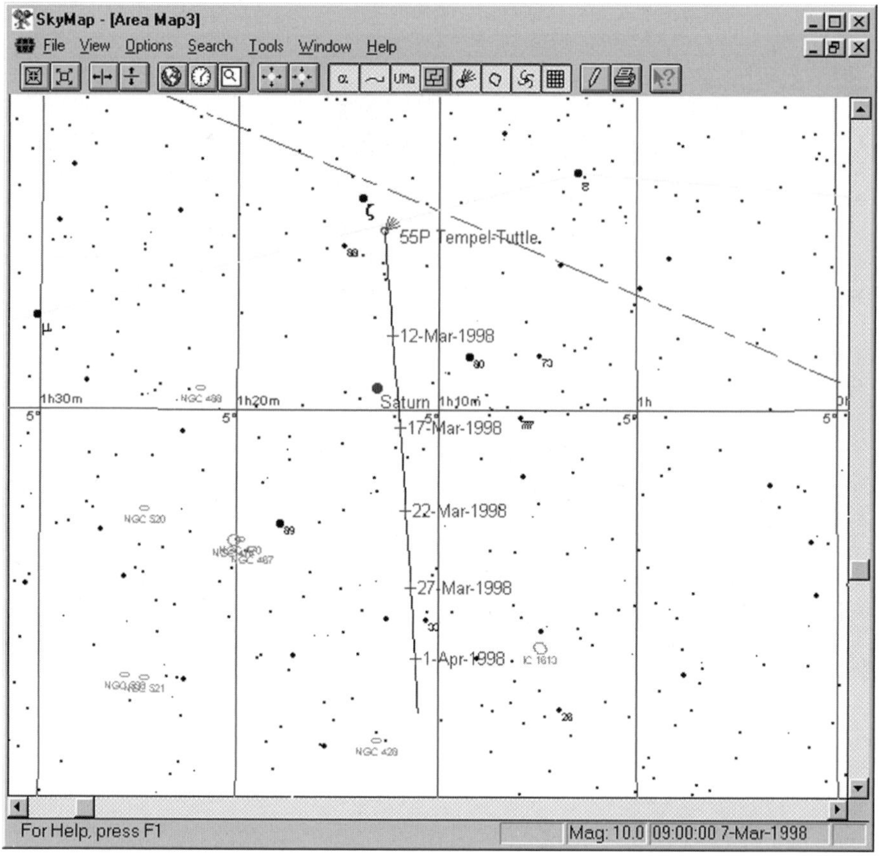

You can even plot field of view circles for various eyepiece/telescope combinations. In my opinion one of the best astronomical software packages available!

Figure 2.5. SkyMap "track" option for Comet Tempel-Tuttle.

Cybersky

Where SkyMap concentrates on mapping and charting, Cybersky's emphasis is more at the personal planetarium level. It too displays the stars, constellations, deep sky objects, planets etc. However, its main attractions are in its educational features and, in particular, its ability to animate displays allowing you to see "movies" of the changing appearance of the sky over time. Cybersky is shareware and its use is limited to 30 days but not all features of the program are available in this version. Versions for both Windows 3.1 and Windows95/98 are included on the CD-ROM. Cybersky

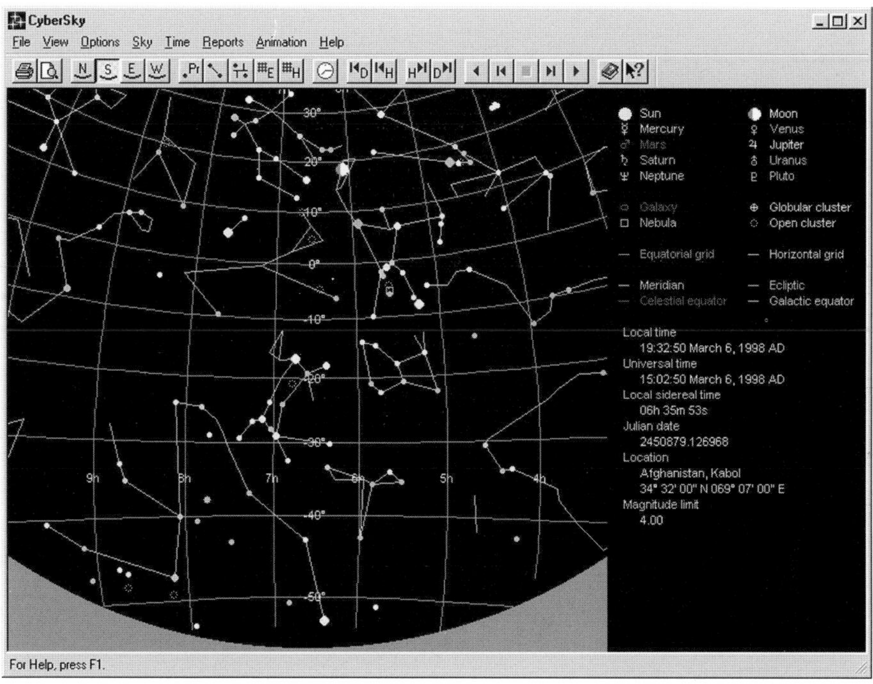

Figure 2.6. Cybersky "horizon" view – not evident here is the fact that the stars are plotted in colour.

too has a Homepage where you can contact the author and find out the latest news.

Not evident in Figure 2.6 is the fact that stars can be coloured according to their spectral class. A nice touch which considerably adds to the visual impact of the charts, especially when animated. There is an uninstall option with the program which enables you to cleanly remove the program should you so wish. It is a shame all programs don't have this feature.

Included

on CD-ROM

Skyglobe

Skyglobe is an easy to use planetarium program that runs under DOS. It supports a mouse and display resolutions ranging from the old CGA standard up to SVGA ones. It even performs well on a 286. Why bother with a program that runs under old fashioned DOS? Well the answer is that many old DOS machines are now available at car boot sales for next to nothing. You can therefore, for very little outlay, have a dedicated planetarium PC for your observatory. Just like the professionals do!

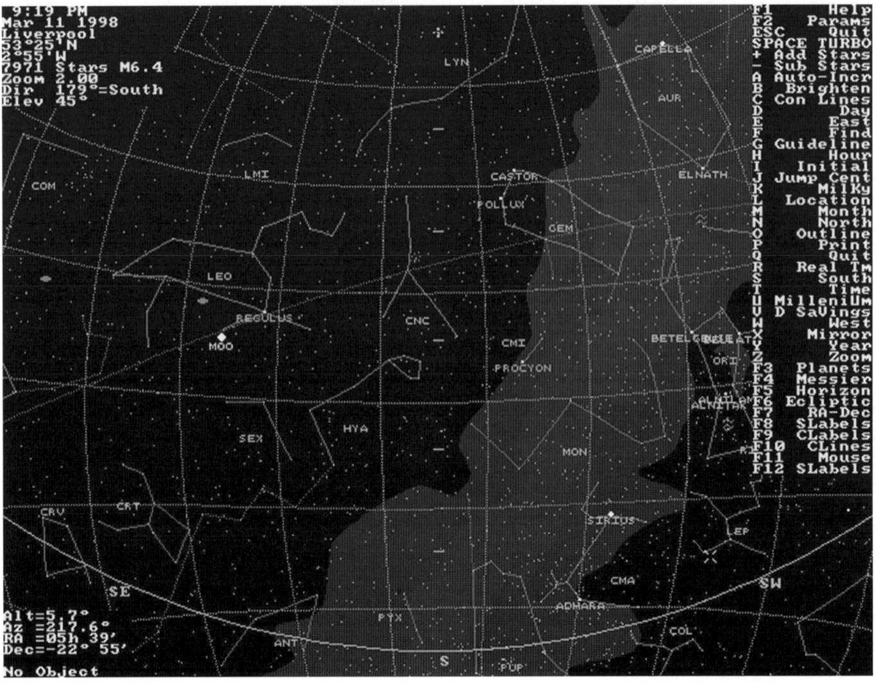

Figure 2.7. Skyglobe – Planetarium program for DOS.

Skyglobe can be set to show the heavens as they would appear from one of more than 200 cities around the world, and once set-up will retain the location setting each time it is run. Most functions are controlled by a single keystroke, many of which can be reversed by using the shift key in conjunction with the same key; i.e. "Z" for zoom in, "shift+Z" to zoom out. Particularly impressive is the rapid panning around the sky using the mouse.

Version 3.6 has 29,000 stars, 110 Messier objects and the "best of the NGC" comprising 112 objects. Other features, not really applicable to the observatory role, include showing the movement of the heavens in real time or speeding it up to watch the sky traverse an entire evening, month, year, or millennium, all in the matter of a few minutes. Version 4.0 is now out which shows many improvements but, in my experience, was "shaky" under Windows95.

Cosmohood

Written by Jeff Bondono, this shareware program adds a new twist to sky simulations. It is one of my

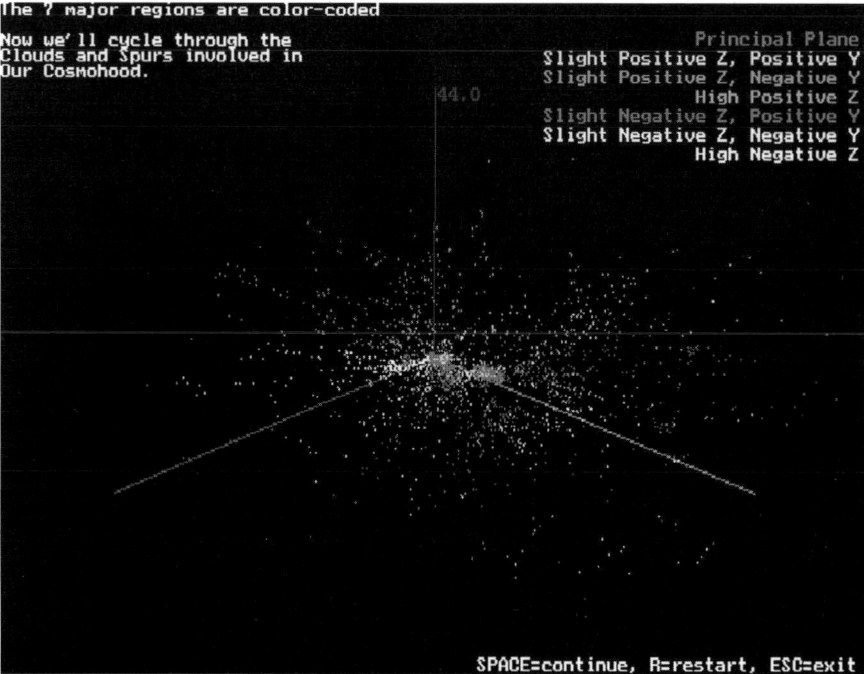

Figure 2.8. Cosmohood 3-D perspective of nearby galaxies.

favourites and the emphasis is very much on the place of the Milky Way galaxy in our cosmic neighbourhood – hence the name. By means of 3-D animations we successively view our place relative to surrounding galaxies, from the local group out to several distant galaxy clusters. It includes the 2,368 nearest galaxies from the Nearby Galaxies Catalogue by Tully. The galaxies are plotted in 3 dimensions and the whole image rotates to enable their relative positions to be appreciated. You can even build your own animations using a scripting language. It runs under DOS and is Shareware. A novel program which I am sure you will want to keep.

Skytimes

This is one of the simplest but most useful little programs around. It is from the SkyMap people and, for any place on any day, it will display the rising and setting times for sun and moon plus all the planets. Twilight times are also given so you will know when you can begin your observing session. It is probably the fastest way to find out what is happening on a particular night. If this is all you need to know then it is

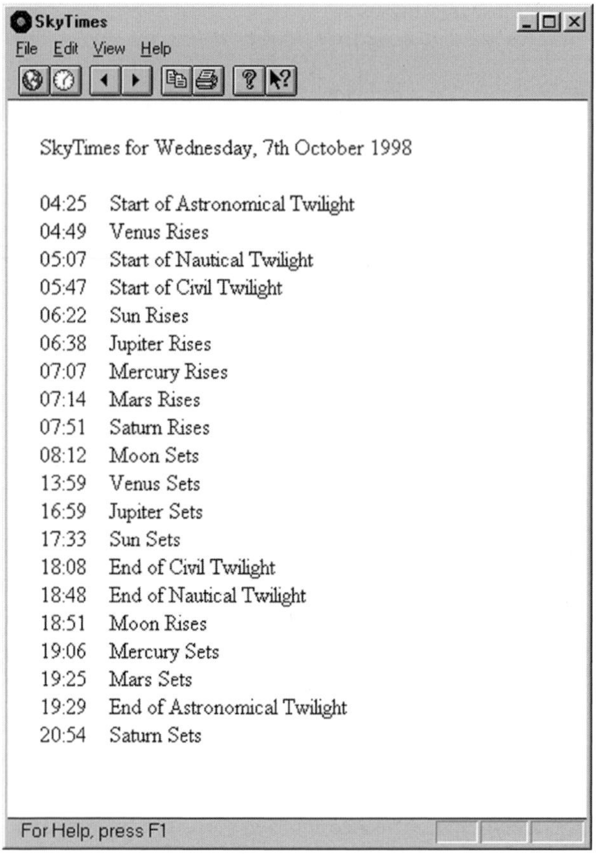

SkyTimes

File Edit View Help

SkyTimes for Wednesday, 7th October 1998

04:25	Start of Astronomical Twilight
04:49	Venus Rises
05:07	Start of Nautical Twilight
05:47	Start of Civil Twilight
06:22	Sun Rises
06:38	Jupiter Rises
07:07	Mercury Rises
07:14	Mars Rises
07:51	Saturn Rises
08:12	Moon Sets
13:59	Venus Sets
16:59	Jupiter Sets
17:33	Sun Sets
18:08	End of Civil Twilight
18:48	End of Nautical Twilight
18:51	Moon Rises
19:06	Mercury Sets
19:25	Mars Sets
19:29	End of Astronomical Twilight
20:54	Saturn Sets

For Help, press F1

Figure 2.9. Skytimes – Twilight start and end plus rising and setting times for a variety of objects.

much quicker than using a planetarium program. It is one you will keep using over and over again. It is shareware and restricted to 30 days.

URLs Featured in this Chapter

SkyMap & Skytimes Homepage:
http://www.skymap.com

Cybersky Homepage:
http://cybersky.simplenet.com

Cosmohood Homepage:
http://bondono.hypermart.net/

Useful Sources of Software

University of Arizona, Students for the Exploration & Development of Space (SEDS):
ftp://ftp.seds.org/pub/software/pc/stars

Greg Roberts Homepage:
http://da.saao.ac.za/~grr/download.html

Birmingham Astronomical Society:
http://www.mee.aston.ac.uk/astro/software.htm

Other Planetarium Programs' Homepages

There are many many more but the following I would recommend as worth checking out.

Deep Space by David Chandler
http://www.csz.com/dschandler
Shareware version available.

TheSKY by Software Bisque
http://www.bisque.com
Market leader but no shareware version.

Graystel Star Atlas
http://members.aol.com/graystel
Excellent UK program – demo only available.

Earth Centred Universe
http://fox.nstn.ca/~ecu/ecu.html
Shareware version available.

Mystars by David Patte
http://www.relativedata.com
Shareware version for Windows available.

Guide Star Chart Homepage by Project Pluto
http://www.projectpluto.com
Sample printouts available.

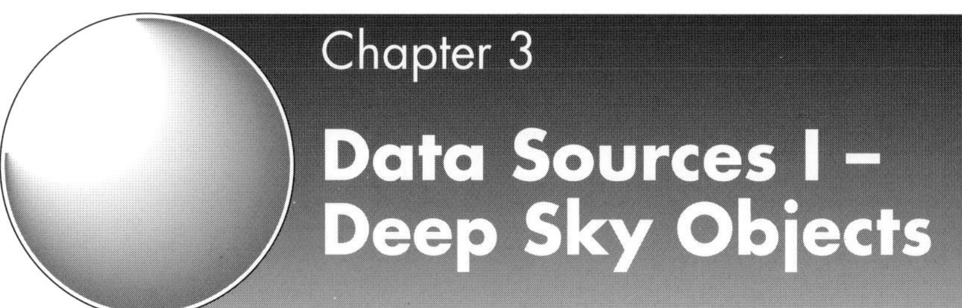

Chapter 3
Data Sources I – Deep Sky Objects

Background

In Chapter 2 we looked at various software packages which displayed the night sky, its stars, planets and deep sky objects. These are in fact mathematical models which simulate the sky. In order to do this, these programs incorporate a variety of astronomical catalogues. Many of these catalogues are available via the Internet but first a word of warning. Every astronomical catalogue has errors. Even Messier's catalogue of only 110 objects has one – the entry M102 is a duplicate of M101. John Flamsteed's star catalogue recorded the planet Uranus (91 years before its discovery!) as the star 34 Tauri. The same is true for more modern catalogues, only more so because of their tens of thousands of entries. Even the Hubble Guide Star Catalogue, which you might assume to be perfect, has errors. These are due, in the main, to the automatic scanning process used which was confused by such things as plate defects, asteroid trails and unresolved stars. So when you discover something unusual, before you announce your discovery to the world, bear in mind the accuracy (or lack of it) in the data against which you are checking. Equally it is unfair to blame the software writer for these errors – they are doing their best with what is available.

The NGC/IC Project

Undoubtedly the best known catalogue of deep sky objects, after Messier's that is, is the New General Catalogue (NGC) and its supplements the Index Catalogues (IC). New is a relative term here. It was published by J.L.E. Dreyer in 1888 as his attempt to collect, in one single catalogue, all the nebulae (including galaxies which were then classified as nebulae) and star clusters. The IC additions came in 1895 and 1908. Dreyer compiled the NGC from a variety of lists and catalogues produced by a range of observers including the Herschels. Because of differences in the telescopes used, observing locations and their abilities, it is no surprise that the accuracy of the data varies hugely. Many NGC/IC numbers have been found to refer to stars and even blank parts of the sky. There are also clear cases of duplication. There have been attempts to correct the worst errors over the years with the Revised NGC (RNGC) of 1973 and NGC2000 in 1988 but these only began the process. The NGC/IC is an attempt by a team of amateur and professional astronomers to correctly identify all the NGC/IC objects and correct that data where necessary.

Their ambitious project has no completion date and is a bit like the painting of the Forth Bridge, it possibly never will be finished. Such is the magnitude of what they are attempting. Each revision just makes it a little more accurate than the last one, with perfection always in the future.

To correct the catalogue they are collecting images for each object and assembling data such as position, classification, magnitude, diameter, colour etc. They have taken the decision to publish their results electronically on the Internet at their Homepage (Figure 3.1, *opposite*). These are very much early days but much has already been done. Feedback is actively encouraged and by doing so you will be helping to get the catalogue "cleaner" and have some fun in the process! Try the Puzzles Section (one is included on the CD-ROM) if you wish to get your teeth into something really interesting!

If you have a query on a particular object then there is a good chance that they will have some corrected data for it already. To check it out you can either look at tables produced by four team members (Dr. Harold G. Corwin, Wolfgang Steinicke, Malcolm Thomson and

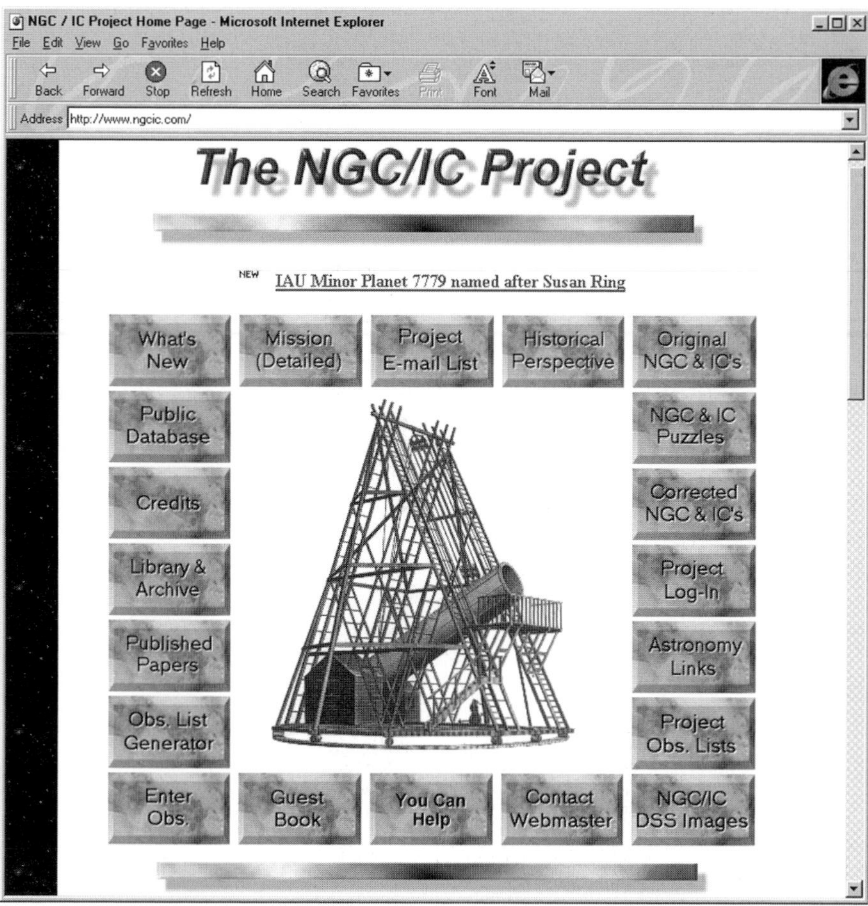

Figure 3.1. The NGC/IC Homepage – note the picture of William Herschel's "40 foot" telescope, indicating how "new" the NGC is.

Steve Gottlieb) or query the Public Access Database (Figure 3.2, *overleaf*).

The Astronomical Data Center

The definitive source of data for professional astronomers is the Astronomical Data Center (ADC) at NASA's Goddard Space Flight Center. It specialises in archiving and distributing collections of data which have been published by professional astronomers. Around 860 astronomical data-sets and 1500 journal tables are stored here. These data-sets are in digital computer readable format. These catalogues, to give

Figure 3.2. Public Access Database – query for NGC6/NGC20.

them their correct title, are available for downloading from ADC's public FTP (file transfer protocol) site. However due to the size of many of these catalogues much of the data is also available on CD-ROM. You can imagine the size of your telephone bill if you were to download the Hubble Guide Star Catalogue!

To help scientists find data-sets, the WWW site has powerful search facilities which interrogates the text "readme" files accompanying each catalogue. The site also has software which can be downloaded and used for accessing the catalogues. This software is available for a variety of hardware platforms. Be aware that, with the exception of the "readme" files, the data is in compressed Unix format. The CD-ROM versions of selected data is available in PC ASCII (text) format. On the Homepage some data is made available for amateur

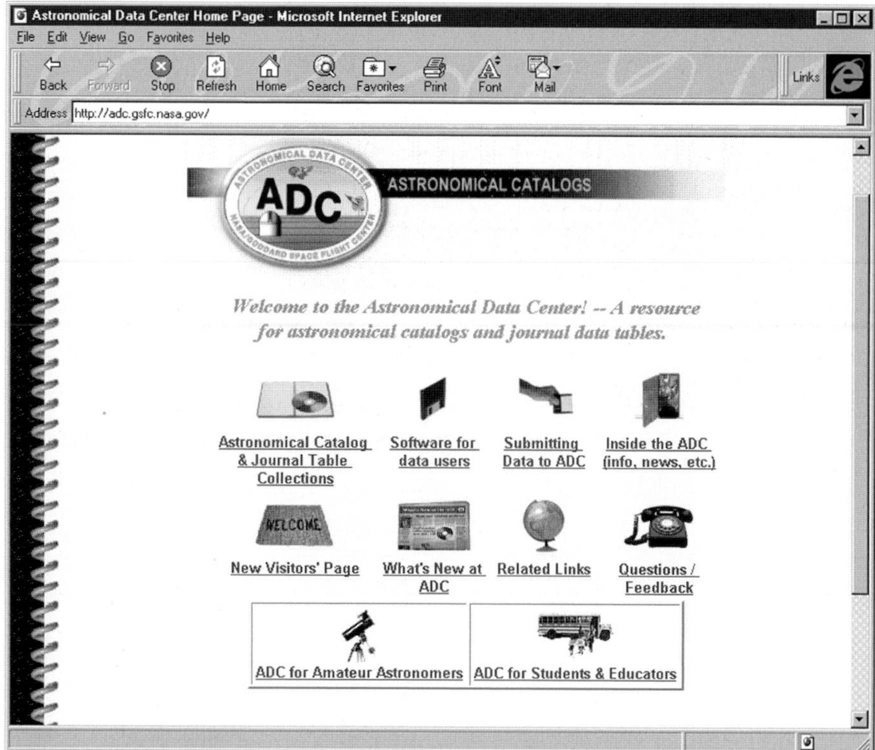

Figure 3.3. The Astronomical Data Center Homepage.

astronomers – assuming that the Unix versions are unsuitable (see Figure 3.4, *overleaf*).

If you require more information about the ADC and its data-sets, the best place to start is their Frequently Asked Questions page which is included on the accompanying CD-ROM to this book. It also offers some excellent guidance on the definitive position regarding star names. No doubt you have been asked by non-astronomers about naming a star after a loved-one. You will find here all you need to know in order to steer them clear of this dubious operation.

The Saguaro Astronomy Club

(SAC) Databases

So far we have looked at two specific sources for data, the NGC/IC Project and the ADC. However, for the

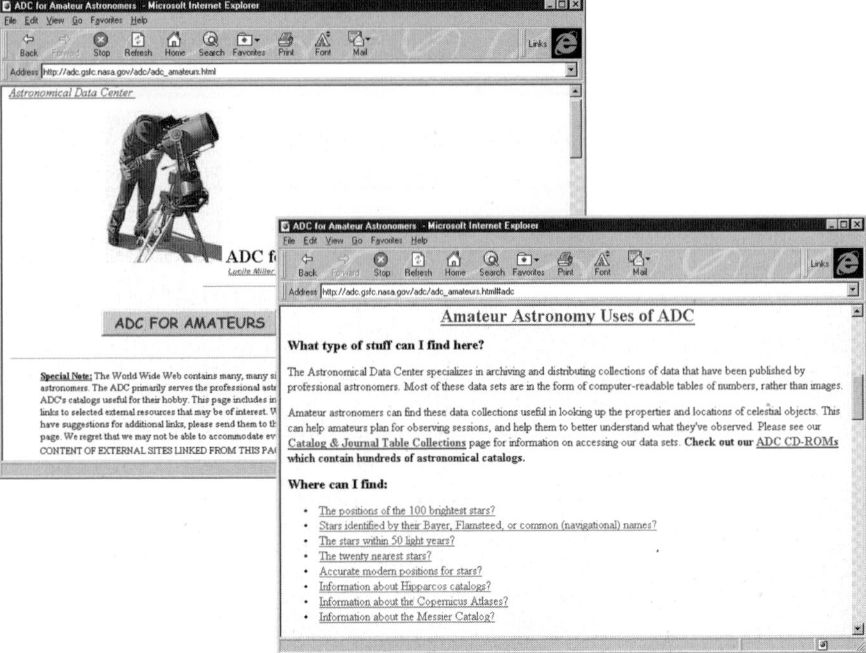

Figure 3.4. Sources of data for the amateur astronomer.

amateur deep sky observer a comprehensive single source, where data from relevant catalogues has been drawn together, would probably meet his needs more effectively. For example, he or she would typically want to know what was visible tonight and probably by constellation. If it is galaxies which are the particular target then whether the galaxy is an NGC/IC object, an UGC (Uppsala General Catalogue) one, or an MCG (Morphological Catalogue of Galaxies) one, is probably irrelevant. The same is true for nebulae, clusters, supernova remnants or even quasars. Fortunately, thanks to the Saguaro Astronomy Club (SAC) of Phoenix, Arizona, there is one such single source available.

The Saguaro (pronounced sa-war-oh) has two main databases and 16 extract files which feature the "best of selections" for a variety of interests and needs, plus a few other well-known catalogues. The two main databases are the Deep Sky Database (Version 6.2) and the Double Star Database (Version 2.1). The former contains information such as size, magnitude, type etc., for over 10,000 clusters, galaxies, and all types of nebulae. The latter contains details of over 10,000 multiple star systems. Thanks to their generosity, these databases are included on the accompanying CD-ROM and have already been unzipped so are

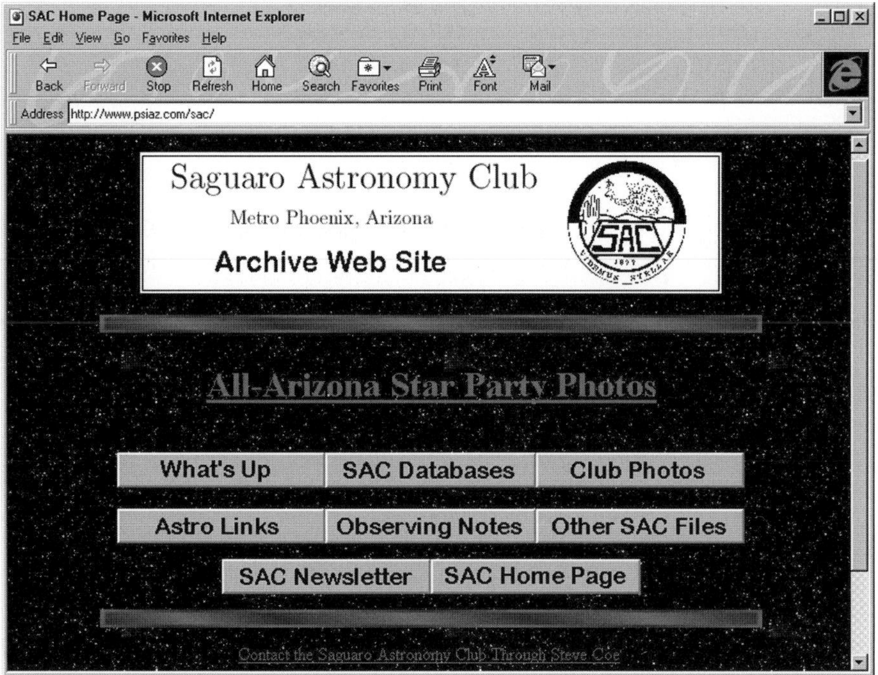

Figure 3.5. The Homepage for the Saguaro Astronomy Club and their extensive collection of Databases.

ready to copy straight to your hard drive and start running.

Much effort by the club (virtually all the Database Team are club members) has gone into assembling these databases. As they have been compiled by amateur astronomers for amateur astronomers, they include the type of data we need rather than swamping us with irrelevancies. For instance, each object is cross-referenced to which page/chart it appears on in both the Uranometria Star Atlas and Tirion's Sky Atlas 2000. A sensible idea. Again they welcome feedback (contact Steve Coe) and should you find errors or omissions in the data, please report these back so that the next version will be nearer perfection.

Included on CD-ROM

Essential to all databases is a means to access the data. Here SAC comes to our rescue again by providing a report generator, SACREP (Figure 3.6, *overleaf*). Written by A.J. Crayon and Dan Ward, this DOS based program is a typical no frills dBase system which nevertheless gets the job done. I will describe a typical query using SACREP on the Deep Sky database but the modus operandi for the Double Star database, using the report generator SACDBL, is identical. First however you will need to copy the SACDS sub-directory to

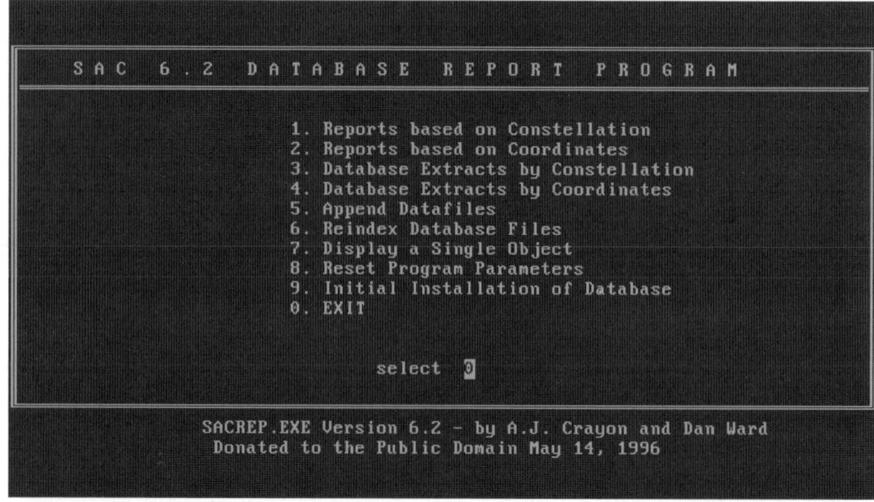

Figure 3.6. Opening menu screen for the report generator, SACREP.

your hard drive. When you first run the program you must select item 9 on the main menu. This is to initialise the database and set the correct path to the

Figure 3.7. Query for all galaxies brighter than magnitude 13 in Andromeda.

```
        SAC 6.1 DATABASE
CONSTELLATION REPORT GENERATOR
------------------------------

ENTER CONSTELLATION ABBREVIATION    AND

ENTER BRIGHTEST MAGNITUDE           -1.0
ENTER DIMMEST MAGNITUDE             13.0

PRINTER, DISK? (P,D)                D

ENTER TYPE OF OBJECTS               GALXY

REPORT FORMAT  (Column Width)
  1=80 2=128 3=160 4=185 5=245      1


ENTER TO CONTINUE OR 0 TO QUIT


      ENTER FILENAME.EXT    Andgals.txt_
```

directory. Now, suppose we want to find all the galaxies brighter than magnitude 13 in the constellation of Andromeda. This is option 1 on the main menu.

Much of the data is in coded form so the file "Sacdoc.txt" which explains this needs to be first printed out. From this we learn than the abbreviation for Andromeda is "AND" and the code for galaxies is "GALXY". We can now fill the query form in (Figure 3.7) using the downarrow key to move to the next field. We have options for printing directly or producing a (text) file in various widths. In the example I have opted for a text file so I have been prompted to give a file name.

Below is just the first few lines of the report generated. The full report runs to 2 pages. So there are more galaxies than just M31 in Andromeda! Note the chart references for Uranometria and Sky Atlas 2000.

If for example you opt for creating a file for loading into a wordprocessor it is as well to be aware that, for columns to line up, you will need to select a uniformly spaced font such as Courier, otherwise the resultant file will be a jumble.

Table 3.1.

Page No. 1 Viewing List For AND Mag -1 To 13
03/16/98

Saguaro Astronomy Club (SAC)
Deep Sky Database Version 6.2
Report Format 1

OBJECT NAME	RIGHT ASC	DEC	SKY ATLAS 2000	U2000	TYPE	MAG	SIZE	CON
NGC 7831	00 07.3	+32 35	04	089	GALXY	12.8	1.7'X0.4'	AND
NGC 20	00 09.6	+33 18	04	089	GALXY	13.0	1.7'X1.5'	AND
NGC 29	00 10.8	+33 21	04	089	GALXY	12.7	2'X1'	AND
NGC 43	00 13.0	+30 54	04	089	GALXY	12.6	1.4'X1.3'	AND
NGC 68	00 18.3	+30 04	04	089	GALXY	12.9	2'X1'	AND
NGC 76	00 19.6	+29 56	04	089	GALXY	13.0	1.4'X1.2'	AND
NGC 80	00 21.2	+22 21	04	126	GALXY	12.1	3'X2'	AND
NGC 83	00 21.4	+22 25	04	126	GALXY	12.5	1.3'X1.2'	AND
NGC 97	00 22.4	+29 44	04	090	GALXY	12.3	1.5'X1.3'	AND
NGC 108	00 25.9	+29 13	04	090	GALXY	12.1	2.3'X2.0'	AND
NGC 160	00 36.0	+23 57	04	126	GALXY	12.6	3'X2'	AND
NGC 183	00 38.3	+29 30	04	090	GALXY	12.7	2.1'X1.6'	AND
NGC 205	00 40.4	+41 41	04	060	GALXY	08.1	17'X10'	AND
NGC 214	00 41.4	+25 30	04	126	GALXY	12.3	2'X2'	AND
NGC 221	00 42.8	+40 52	04	060	GALXY	08.1	8'X6'	AND
NGC 224	00 42.8	+41 16	04	060	GALXY	03.4	178'X40'	AND
NGC 233	00 43.4	+30 35	04	090	GALXY	12.4	2.0'X1.7'	AND
NGC 252	00 48.0	+27 37	04	126	GALXY	12.4	1.7'X1.2'	AND

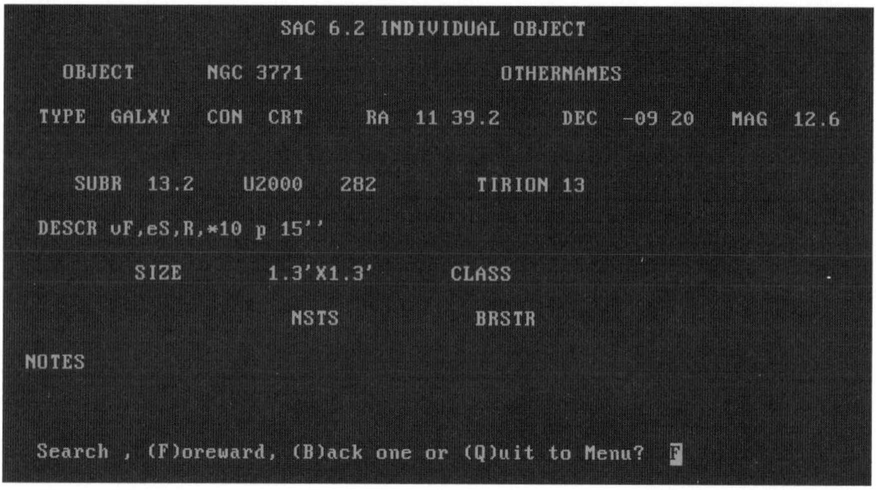

```
                    SAC 6.2 INDIVIDUAL OBJECT

   OBJECT       NGC 3771                    OTHERNAMES

 TYPE  GALXY    CON  CRT      RA  11 39.2      DEC  -09 20    MAG  12.6

    SUBR  13.2     U2000   282          TIRION 13

 DESCR vF,eS,R,*10 p 15''

          SIZE        1.3'X1.3'       CLASS

                        NSTS            BRSTR

NOTES

 Search , (F)oreward, (B)ack one or (Q)uit to Menu? F
```

Figure 3.8. Results from SACREP of a query on a single object, NGC3771.

It is also possible to query on a single object. Figure 3.8 gives a typical return for galaxy NGC 3771. To understand the abbreviations used in the printout you will again need to refer to Sacdoc.txt.

When carrying out a single query it is necessary to be aware of the limitations of SACREP. The data is fixed format, i.e. of fixed field length so for NGC numbers, 8 spaces are allowed with the assumption that the number will be right justified. If we are searching on an NGC object with a number less than 1000, i.e. 3 figures or less, then we could get a nil return. So for example, if searching for NGC 205, two spaces should be entered between the C and the 2. Once having made our selection of an object we can step forward or backward through the database at will.

As mentioned earlier, as well as the two main databases, Saguaro also provide additional sub-selections based on either established catalogues like Messier's or Patrick Moore's Caldwell list, or their own recommendations for a variety of needs (Figure 3.9, *opposite*). Suppose for example your Astronomical Society has a public viewing night, then simply print out the "Public View Session Objects" and you will have plenty of "show-stoppers" to impress and excite your visitors. There are selections for Astrophotographers and Rich Field Viewing (RFT). Remember they are based on the observations of experienced observers and are all the more valuable for that.

Finally just a word about where to find the data. The Deep Sky Database is in sub-directory \SACDS, the Double Star Database in \SACDBL and the additional

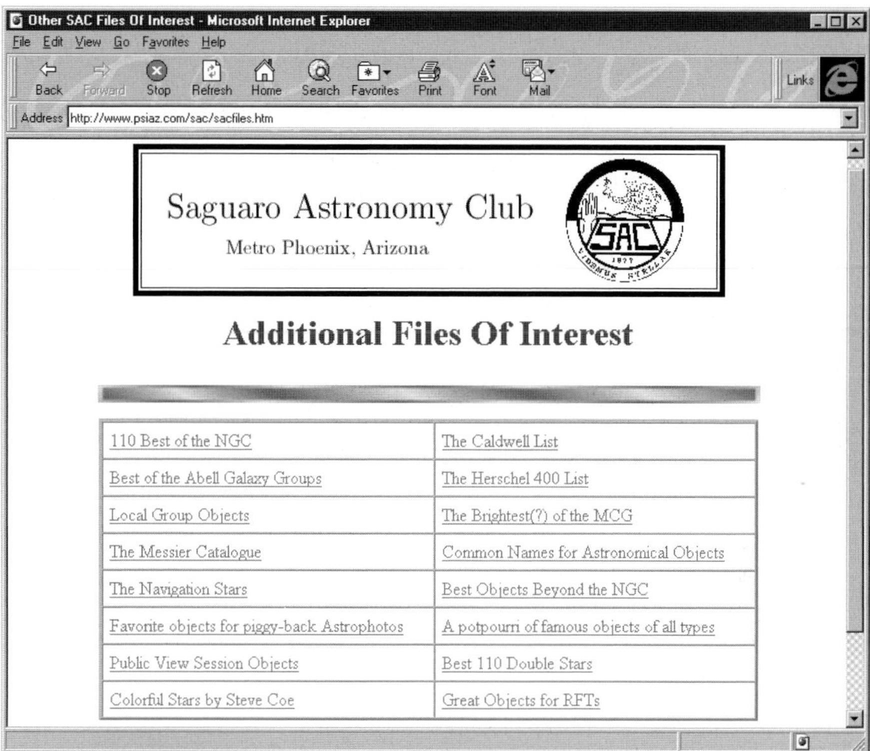

Figure 3.9. The "best of" additional SAC files.

files in \SACFILES. The raw datasets are included for those wishing to perhaps develop their own Report Generators. In addition, Acrobat PDF versions of the complete databases (SACALL.PDF) provide a simpler query facility based on Acrobat's free text search facility (binocular icon). Whichever way you access the data, we all owe a debt of gratitude to the team at the Saguaro Astronomy Club for their work and generosity in sharing it with amateur astronomers everywhere. The true spirit of amateur astronomy is still alive and flourishing in Saguaro.

dObjects

We now look at a commercial Database and Observing Log, dObjects by Jeff Bondono. The version of the software included on the CD-ROM is an evaluation copy only and is restricted to the Messier objects. It does however give a good idea of the capabilities of a modern database with multiple windows, pull-down

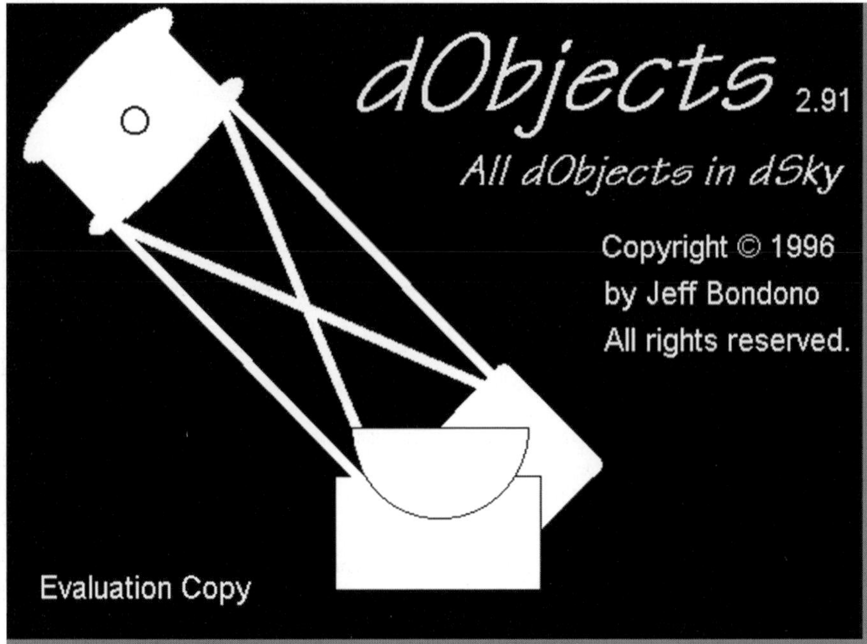

Figure 3.10. dObjects opening screen.

menus, icons and online help. It runs under Windows 95 or higher only.

The full version has more objects than the SAC Database at around 34,000 (not in the evaluation copy remember) but, what makes it stand out to me, are the cross references to articles about the object in popular Astronomical Magazines. How many times have you viewed an object then wished to learn more but could not remember when or where you read an article about it? dObjects solves that. It also includes Internet cross references, again providing the URLs where you can find out more. Figure 3.11 (*opposite*) is a typical screen displaying details of the Ring Nebula M57. Information displayed is that the nebula is magnitude 9.7 with the central star's magnitude at 14.8. The object can be found on Uranometria chart number 117 and there is a plethora of references to astronomical magazines with a "p" indicating that there will be a photograph in the article. Each window can be expanded or reduced to make the information more or less visible. So when you next view a Messier object you will know where to find out some background information.

The second part of dObjects is selecting and compiling viewing lists or logs. The "select objects to view button" does this for us based on any set of criteria such as constellation, brightness type of object etc.. A

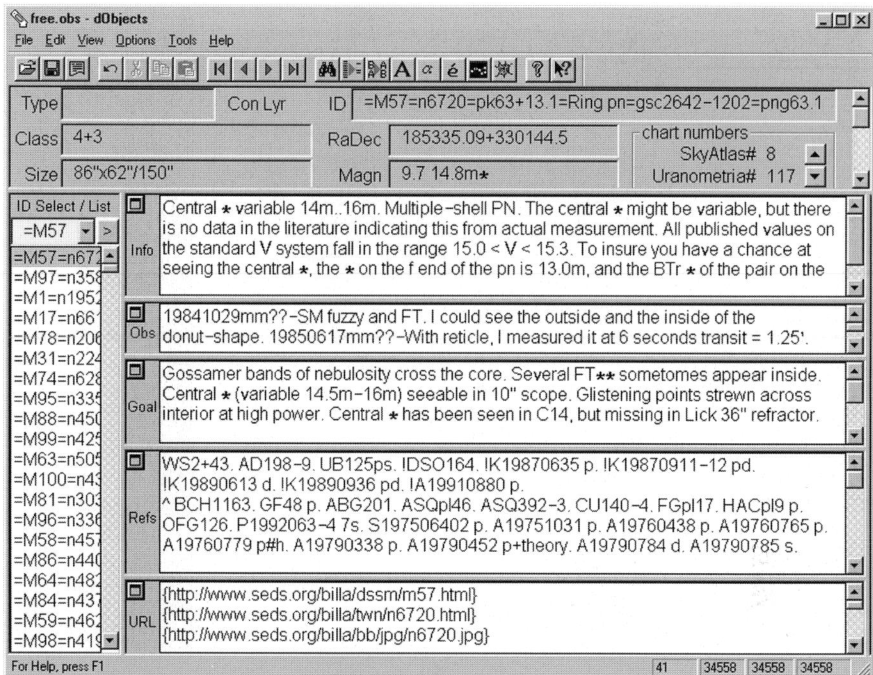

free.obs - dObjects
File Edit View Options Tools Help
Type
Con Lyr
ID =M57=n6720=pk63+13.1=Ring pn=gsc2642-1202=png63.1
Class 4+3
RaDec 185335.09+330144.5
chart numbers
SkyAtlas# 8
Uranometria# 117
Size 86"x62"/150"
Magn 9.7 14.8m*
ID Select / List
=M57
Info
Central * variable 14m..16m. Multiple-shell PN. The central * might be variable, but there is no data in the literature indicating this from actual measurement. All published values on the standard V system fall in the range 15.0 < V < 15.3. To insure you have a chance at seeing the central *, the * on the f end of the pn is 13.0m, and the BTr * of the pair on the
Obs
19841029mm??-SM fuzzy and FT. I could see the outside and the inside of the donut-shape. 19850617mm??-With reticle, I measured it at 6 seconds transit = 1.25'.
Goal
Gossamer bands of nebulosity cross the core. Several FT** sometomes appear inside. Central * (variable 14.5m-16m) seeable in 10" scope. Glistening points strewn across interior at high power. Central * has been seen in C14, but missing in Lick 36" refractor.
Refs
WS2+43. AD198-9. UB125ps. !DSO164. !K19870635 p. !K19870911-12 pd. !K19890613 d. !K19890936 pd. !A19910880 p.
^BCH1163. GF48 p. ABG201. ASQpl46. ASQ392-3. CU140-4. FGpl17. HACpl9 p. OFG126. P1992063-4 7s. S197506402 p. A19751031 p. A19760438 p. A19760765 p. A19760779 p#h. A19790338 p. A19790452 p+theory. A19790784 d. A19790785 s.
URL
{http://www.seds.org/billa/dssm/m57.html}
{http://www.seds.org/billa/twn/n6720.html}
{http://www.seds.org/billa/bb/jpg/n6720.jpg}
For Help, press F1
41
34558
34558
34558

Figure 3.11. dObjects information screen on the Ring Nebula, M57 (NGC6720).

feature of this is the ability to easily refine our criteria and see instantly the new results. dObjects is a sophisticated program and the full system is available to purchase. Further details are available from the Homepage–see below for the URL.

Astrophysical Data Service (ADS) at the Smithsonian Astrophysical Observatory

This NASA funded project at the Smithsonian Astrophysical Observatory in Cambridge, Massachusetts is the place to visit for astronomical abstracts. Most astronomical literature since 1975 is included in the ADS database which is fully searchable online. That runs to around 1 million items! Whilst abstracts are its main feature, ADS also has over 150 astronomical catalogues, some scanned books and impressive data archives.

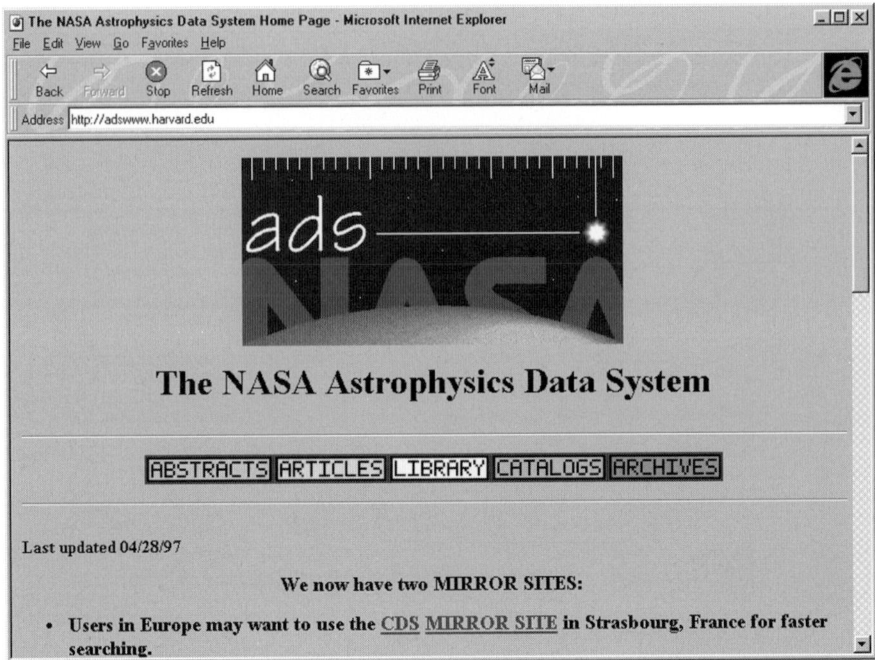

Figure 3.12. The ADS Homepage.

Although perhaps not as user-friendly as some other sites it is nevertheless one to include in your bookmarked web sites.

URLs Featured in this Chapter

The NGC/IC Project
http://www.ngcic.com

The Astronomical Data Center (ADC)
http://adc.gsfc.nasa.gov/adc.html

Saguaro Astronomy Club Homepage
http://www.psiaz.com/sac

dObjects–Jeff Bondono's Homepage
http://bondono.hypermart.net

Astrophysical Data Service (ADS)
http://adswww.harvard.edu

Other URLs not Referred to but Recommended

Arp's Catalogue of Peculiar Galaxies
http://members.aol.com/arpgalaxy

If in doubt always try the "SEDS" pages
http://www.seds.org

National Space Science Data Center
http://nssdc.gsfc.nasa.gov

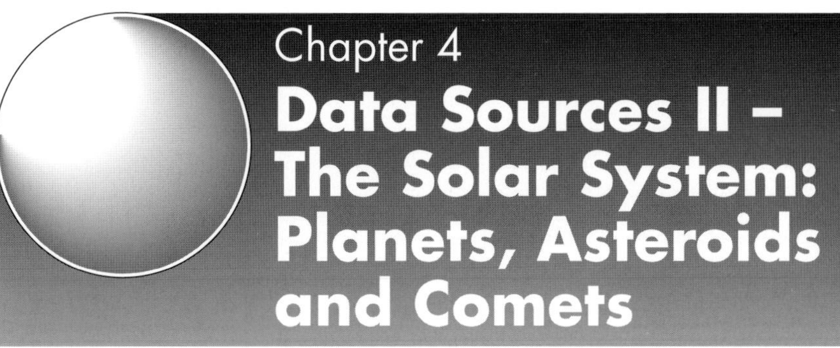

Chapter 4 Data Sources II – The Solar System: Planets, Asteroids and Comets

Background

We now turn to our own solar system and not surprisingly the Internet has an embarrassment of riches on offer. The standard method of finding data on solar system bodies has been the traditional Astronomical or Nautical Almanac and its ephemerides for each year. These feature the (almost) daily tabulated positions and properties of celestial bodies but to the uninitiated they hardly make for enjoyable bedtime reading! To be fair, for the professional or the serious amateur researcher these are still an essential resource but to the ordinary amateur there are now more user-friendly ways of getting the data we want.

Whilst the planets might be regarded as predictable, that is not true of other solar system bodies. For the lesser bodies it pays to expect the unexpected – a new comet or perhaps an asteroid discovery (only likely to be reported in the press if it is going to collide with Earth) is always a distinct possibility. Here the Internet can disseminate new discoveries almost instantly. As referred to several times previously, probably the place to regularly check for new discoveries is the Sky & Telescope Homepage.

I will now turn to just a few of the many excellent sites available. Whilst the Naval Departments on both sides of the Atlantic are the definitive sources of data – you can bet your life on their information being correct

– we will look at ones perhaps more geared to amateur astronomers.

JPL Solar System Dynamics Group

The reason for choosing this site over the Naval Observatories is its fantastic online planetary information service, known as HORIZONS (Figure 4.2, *opposite*). Using this you can output precision positions for the Sun, Moon, planets and nearly all their satellites (well over 50 of them), comets and asteroids. The site has much more and is well worth visiting for its other facilities, such as lists of asteroids visible on a particular night and help on identifying asteroids recorded on images. However it is HORIZONS which makes it stand out and is what I shall concentrate on here.

Figure 4.1. The JPL Solar System Dynamics Group Homepage.

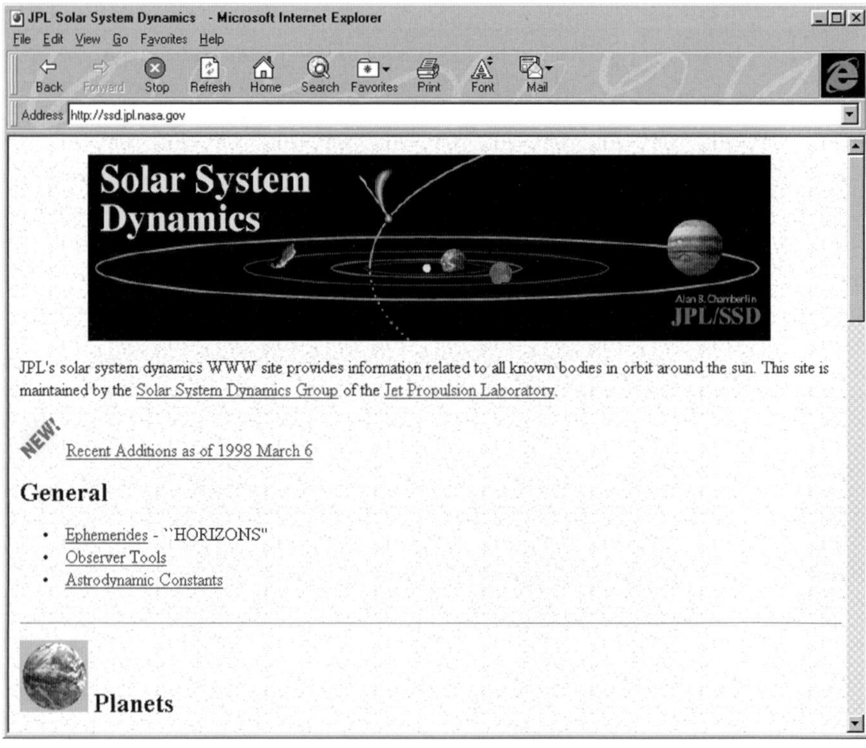

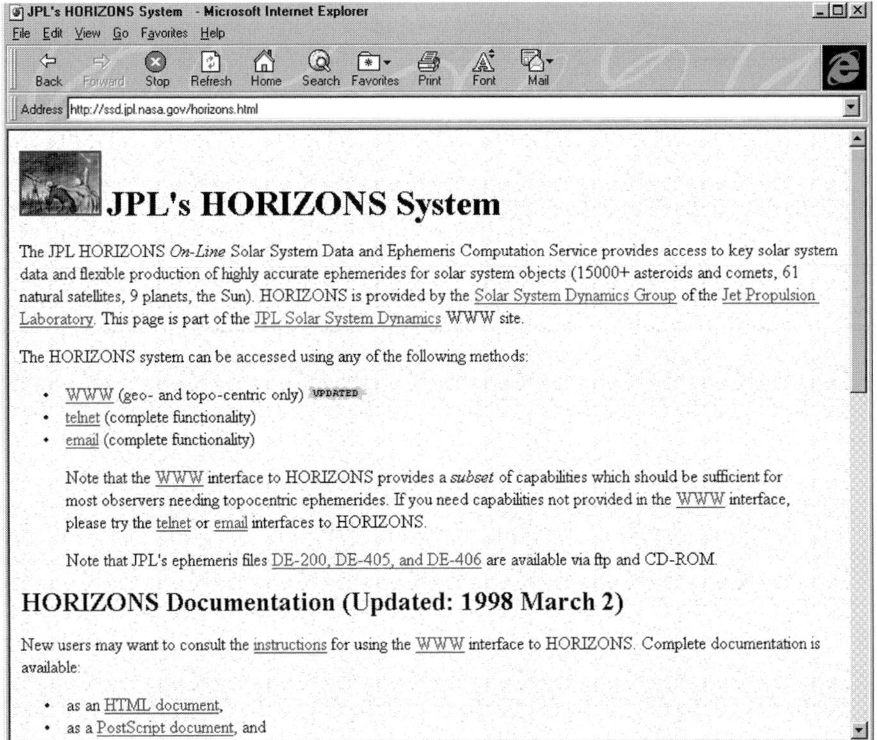

Figure 4.2. The HORIZONS System.

Included

on CD-ROM

JPL publishes a 40 page guide to HORIZONS in both HTML and Postscript format. This excellent guide is worth studying and is included on the accompanying CD-ROM in Acrobat format. It should also be pointed out that HORIZONS can be accessed by simple email and instructions are included in the guide for this method of access, as well as by Telnet (for terminals). What this means is that you can send an email to a machine requesting information. It will extract it for you and duly send it back! What an age we live in. If it is regular repeated requests you are interested in, then this is probably the best way to go.

We will however take a quick look at the WWW browser interface and how to request data. The example shown (Figure 4.3, *overleaf*) is for extracting the ephemeris for the planet Mars. Bodies are split into two categories, Major Bodies (MB) and Small Bodies (SB). The latter is for asteroids and comets, not as you might imagine moons. You need to be able to understand

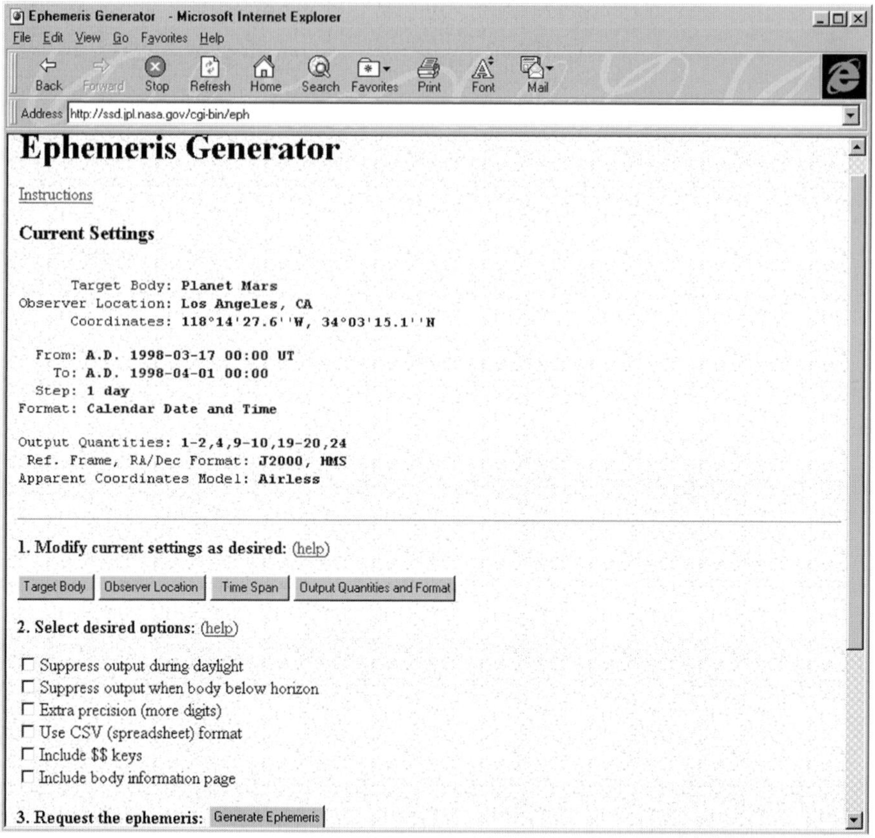

Figure 4.3. Generating an ephemeris for the planet Mars using HORIZONS.

some terminology such as barycentres, as ephemerides can be for the planet centre or the barycentre of the planet plus its moon system. For Mars (or for that matter Mercury or Venus) just use the planet centre. Other information needed is of course your location and obviously the period for which the ephemerides are required plus the time interval required, usually daily.

Other points to mention are that the range of dates covered is 3000BC to 3000AD but remember the solar system does not run like clockwork and any results for dates many years away from today are tentative at best. This has been a very brief introduction to HORIZONS but more information is available in the Guide.

Views of the Solar System

Now for something a little lighter, you will not find any ephemerides here. You might imagine from its name that this site is solely to do with images but whilst it obviously has images aplenty there is much more to it than that. Written by Calvin J. Hamilton, these pages are more a multimedia tour of the solar system – a sort of modern day encyclopaedia with animations, charts, images and descriptive text (Figure 4.4).

It really excels if you want some background information on a particular topic or solar system body. Perhaps you are writing an article for your society magazine and need to know all about Titan for example. This site is probably the best place to start and, in the unlikely event that you need to know even more, links are usually provided onto further webpages

Figure 4.4. The Views of the Solar System Homepage.

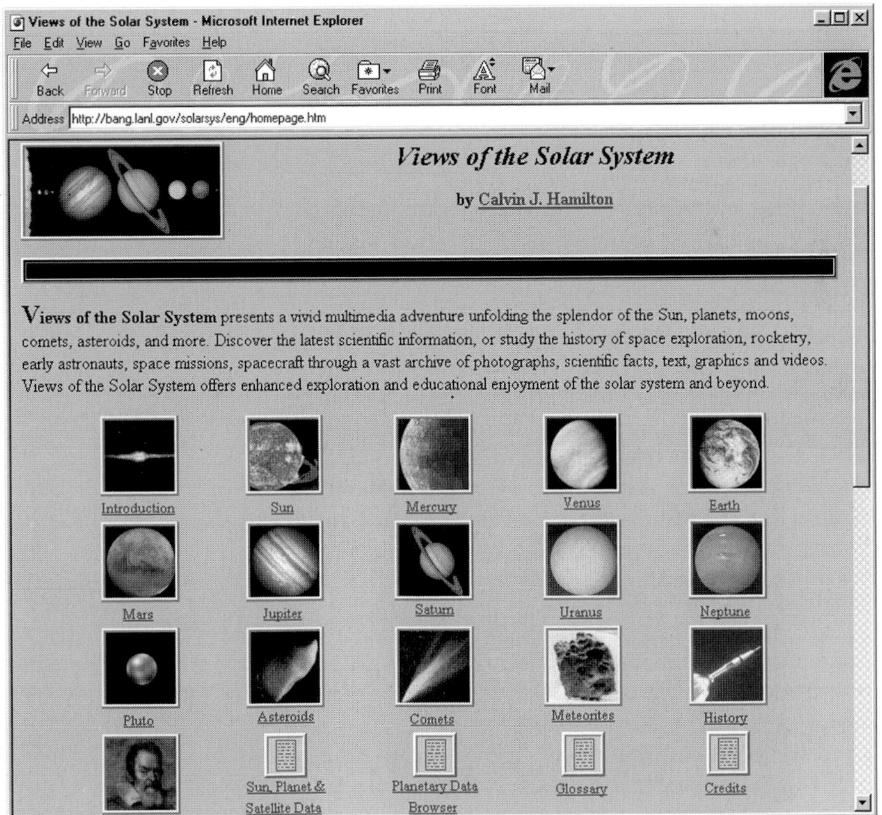

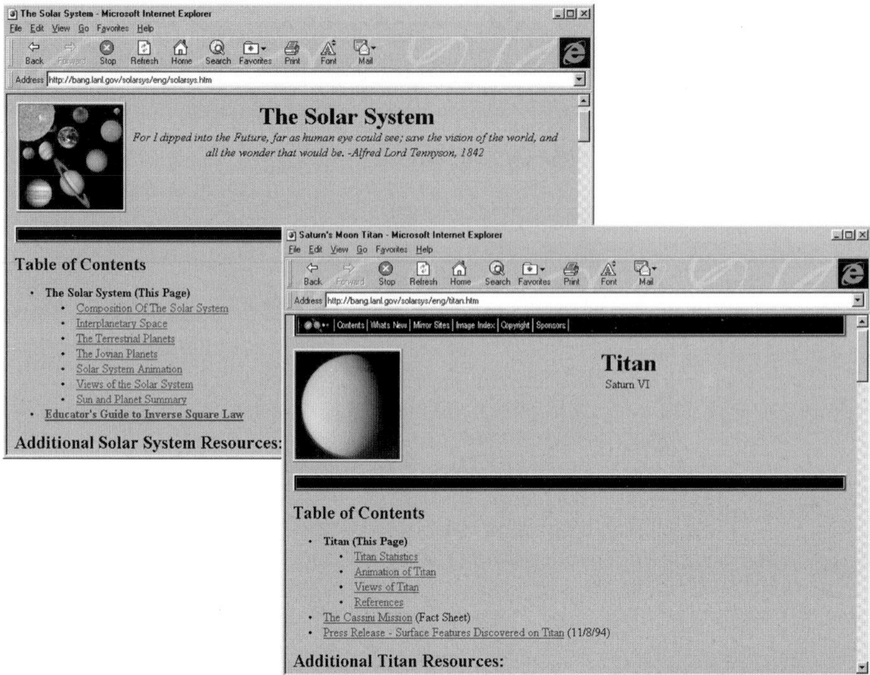

Figure 4.5. Finding information for Titan using Views of the Solar System.

elsewhere. Figure 4.5 gives an idea of how easy it is to find that information for Titan.

Mooncalc

As this book has so far been dominated by American sites, it makes a welcome change to deal with a British one. Not only that, this freeware DOS program written by Dr. Monzur Ahmed is a little cracker! If you are a lunar observer wanting to know when to observe (or an astrophotographer wanting to know when not to) then this little program will tell you everything about the visibility of the Moon. It runs under MS-DOS and is freeware – all the author asks is that you acknowledge the source if publishing the data obtained.

For any place on Earth the lunar position, age, phase, rising and setting times are calculated and displayed in either tables or charts (Figure 4.7, *opposite*). The graphics are impressive too. The Moon can be shown against the stars, just like a planetarium program, either in RA/DEC or ALT/AZ format. A close-up graphic option even includes features such as craters (toggle the C key

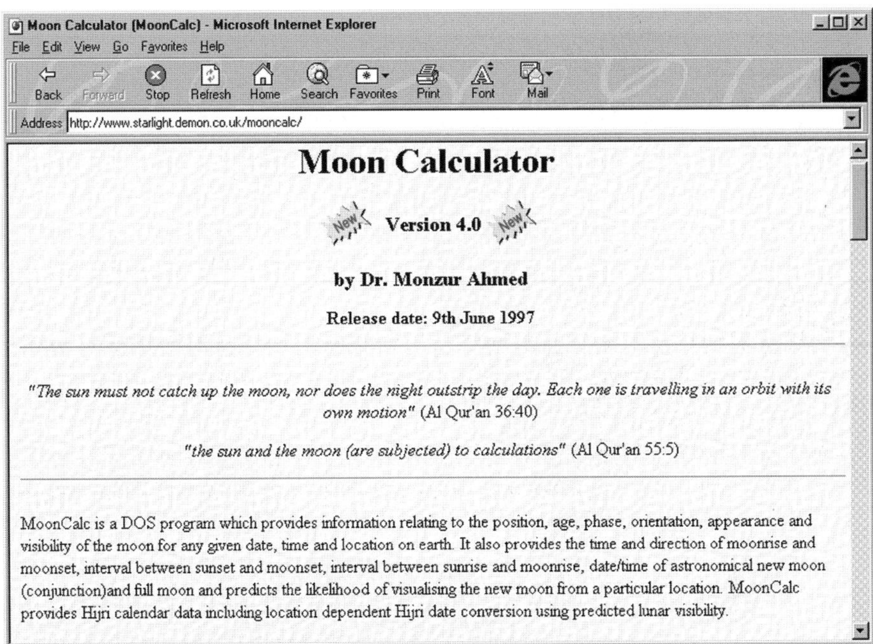

Figure 4.6. The Mooncalc Homepage.

Figure 4.7. Table of lunar data produced by Mooncalc.

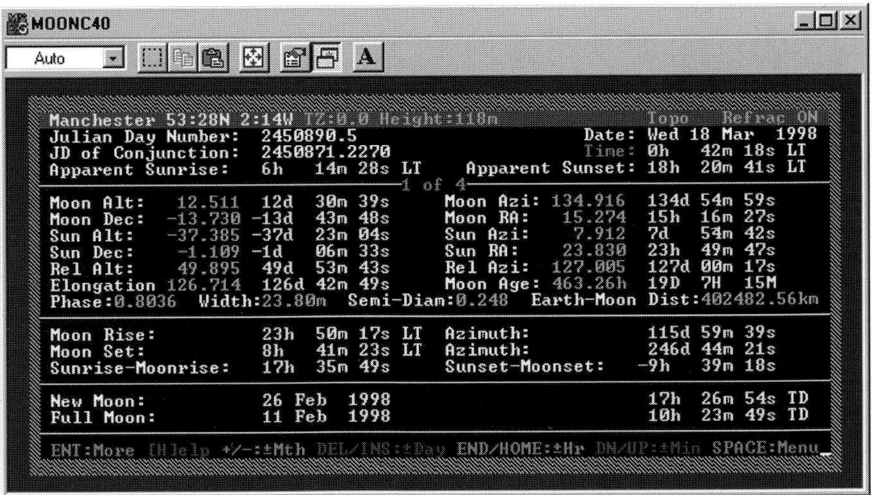

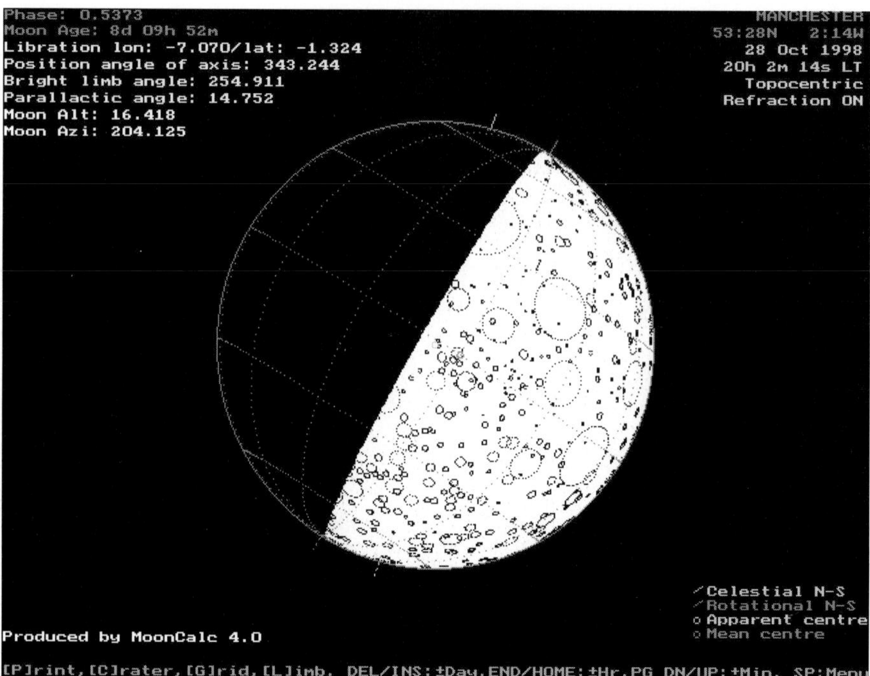

Figure 4.8. Close-up of the Moon showing crater details.

to switch on) and gridlines (G key) – see Figure 4.8. The earliest possible sighting of the Moon is a main feature of the program, which is important to some communities. Again a mixture of graphics and text provide the information (Figure 4.9, *opposite*).

To run the program it will be necessary to copy the directory \Mooncalc to your hard-drive as the program needs to write a file with your particular defaults, such as your location. All in all an excellent program you will keep returning to.

Minor Planet Center (MPC)

Despite what you would expect from its name, this official site is the definitive source of information on not just minor planets (asteroids) but comets as well. A very prophetic decision, as in some recent instances the distinction between the two is not as straightforward as previously imagined. The Minor Planet Center (MPC) operates at the Smithsonian

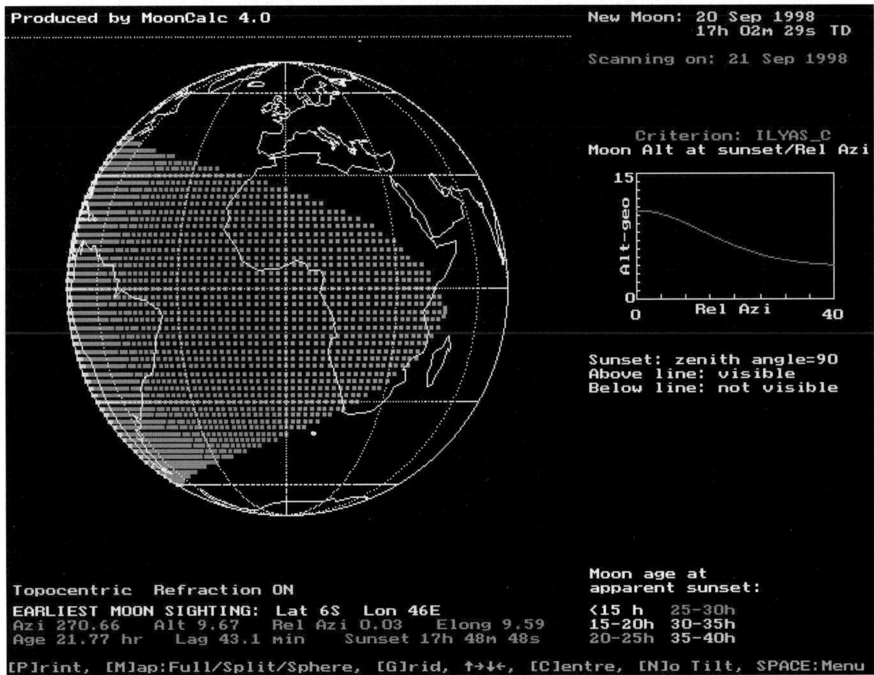

Figure 4.9. Mooncalc chart showing the visibility of the new Moon plotted on a map of the world.

Astrophysical Observatory under the auspices of Commission 20 of the International Astronomical Union (IAU). It is charged with the efficient collection, computation, checking and dissemination of astrometric observations and orbits for minor planets and comets, via the monthly Minor Planet Circulars and the Minor Planet Electronic Circulars, which are issued as necessary.

Their definitive publication, The Minor Planet Circulars (or more appropriately it should be called the Minor Planets & Comets), is available only by subscription. Details of how to subscribe are available on the web site and several sample back issues are available for downloading to give an idea of just what you would get. It also publishes an annual Catalogue of Cometary Orbits. There is however much data on the site which is freely available.

Probably the most impressive is the Minor Planet Checker (Figure 4.11, *overleaf*). Using this online entry form you can quickly check if an unidentified object on one of your photographs or images is a minor planet or not. Simply enter the date plus either the co-ordinates of the area in question (or the NGC number if a galaxy), a radius of search and a limiting

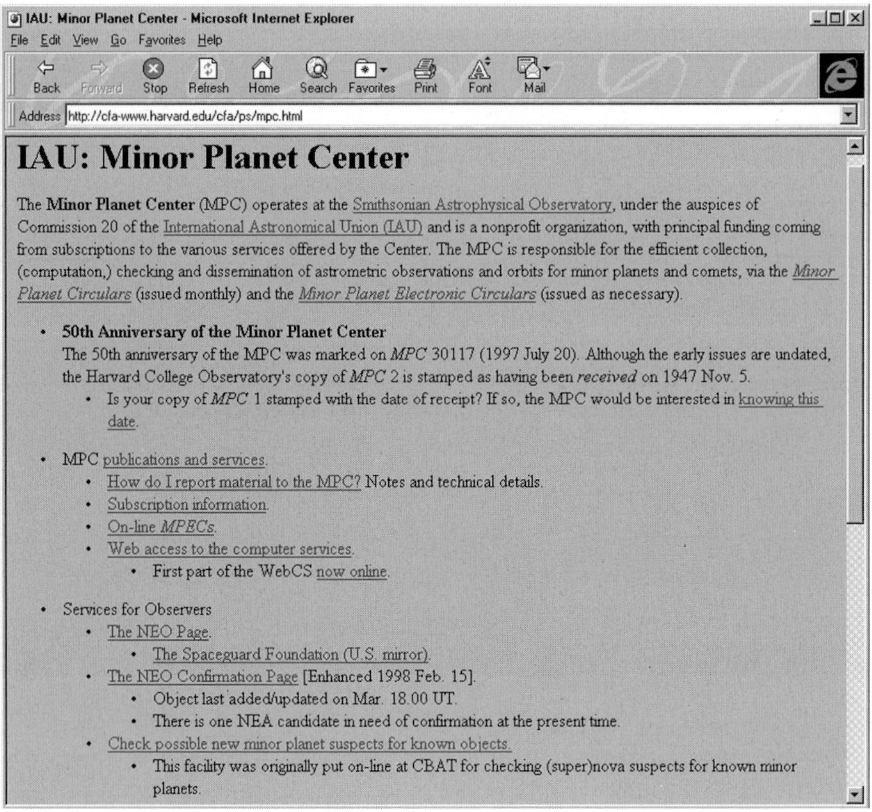

IAU: Minor Planet Center - Microsoft Internet Explorer

File Edit View Go Favorites Help

Back Forward Stop Refresh Home Search Favorites Print Font Mail

Address http://cfa-www.harvard.edu/cfa/ps/mpc.html

IAU: Minor Planet Center

The **Minor Planet Center** (MPC) operates at the Smithsonian Astrophysical Observatory, under the auspices of Commission 20 of the International Astronomical Union (IAU) and is a nonprofit organization, with principal funding coming from subscriptions to the various services offered by the Center. The MPC is responsible for the efficient collection, (computation,) checking and dissemination of astrometric observations and orbits for minor planets and comets, via the *Minor Planet Circulars* (issued monthly) and the *Minor Planet Electronic Circulars* (issued as necessary).

- **50th Anniversary of the Minor Planet Center**
  The 50th anniversary of the MPC was marked on *MPC* 30117 (1997 July 20). Although the early issues are undated, the Harvard College Observatory's copy of *MPC* 2 is stamped as having been *received* on 1947 Nov. 5.
  - Is your copy of *MPC* 1 stamped with the date of receipt? If so, the MPC would be interested in knowing this date.
- MPC publications and services.
  - How do I report material to the MPC? Notes and technical details.
  - Subscription information.
  - On-line *MPECs*.
  - Web access to the computer services.
    - First part of the WebCS now online.
- Services for Observers
  - The NEO Page.
    - The Spaceguard Foundation (U.S. mirror).
  - The NEO Confirmation Page [Enhanced 1998 Feb. 15].
    - Object last added/updated on Mar. 18.00 UT.
    - There is one NEA candidate in need of confirmation at the present time.
  - Check possible new minor planet suspects for known objects.
    - This facility was originally put on-line at CBAT for checking (super)nova suspects for known minor planets.

Figure 4.10. The Minor Planet Center (MPC) Homepage.

magnitude. A list will be returned with all known asteroids and comets in the search area. So if you do think you've discovered something, this is the place to start before making any claims. If more amateurs did so then many false claims would not see the light of day!

In addition the MPC makes freely available on its web site what it terms, Lists & Plots, for both minor planets and comets. These include much data on topical subjects such as "near earth objects" for both past ones and those due in the next 25 years (for at least the ones we know about). In the case of comets, recent comet magnitude tables are also available together with lists such as periodic comets awaiting their first return. Several of these are included on the CD-ROM.

Minor Planet Checker - Microsoft Internet Explorer

File Edit View Go Favorites Help

Back Forward Stop Refresh Home Search Favorites Print Font Mail

Address http://cfaps8.harvard.edu/~cgi/CheckSN.COM

Minor Planet Checker

Use the form below to prepare a list of known minor planets in a specified region. This feature was originally intended for checking possible (super)nova candidates, hence the prominence given to searches around a specific galaxy. Notes on using this form are given at the bottom of this page.

If you wish to report the non-functioning of (or errors in) this service, please use this feedback form.

Produce list | Clear/reset form

Date : 1998 03 18.49 UT

Produce list of known minor planets around:

this J2000.0 position:
R.A. = Decl. =

this galaxy : NGC 5000

Radius of search = 15 arc-minutes

Limiting magnitude, V = 18.0

Output matches in order of:

increasing distance from specified position

increasing Right Ascension

Figure 4.11. The MPC Minor Planet Checker data entry form.

BAA Comet Section

If you want the latest information on comets then, if you will forgive me for my British bias, the British Astronomical Association's Comet Page is probably the best place to look. It has a latest news section, current comet magnitudes, comet ephemerides, meteor showers and upcoming (incoming?) comets.

For all current comets there are lists of recent observations by amateur astronomers, often with quite modest equipment such as 10×50 mm binoculars. Thus you immediately get an idea of how difficult a particular comet will be to see for yourself. Figure 4.13 (*overleaf*) is just such a description for Comet Hartley2 which was visible with 20×80 mm binoculars.

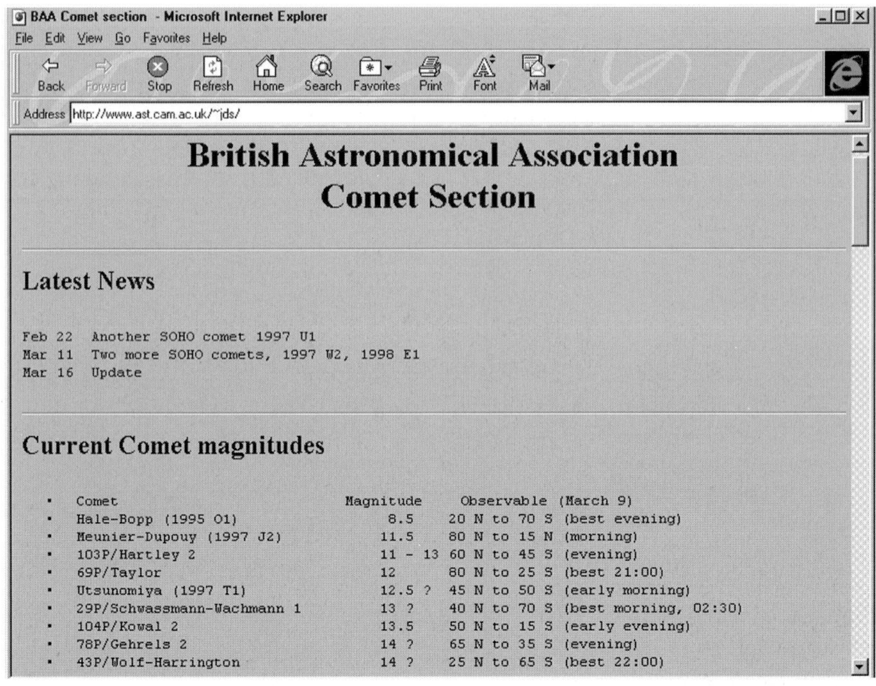
BAA Comet section - Microsoft Internet Explorer

File Edit View Go Favorites Help

Back Forward Stop Refresh Home Search Favorites Print Font Mail

Address http://www.ast.cam.ac.uk/~jds/

British Astronomical Association Comet Section

Latest News

```
Feb 22  Another SOHO comet 1997 U1
Mar 11  Two more SOHO comets, 1997 W2, 1998 E1
Mar 16  Update
```

Current Comet magnitudes

```
•  Comet                          Magnitude    Observable (March 9)
•  Hale-Bopp (1995 O1)               8.5     20 N to 70 S (best evening)
•  Meunier-Dupouy (1997 J2)         11.5     80 N to 15 N (morning)
•  103P/Hartley 2                   11 - 13  60 N to 45 S (evening)
•  69P/Taylor                       12       80 N to 25 S (best 21:00)
•  Utsunomiya (1997 T1)             12.5 ?   45 N to 50 S (early morning)
•  29P/Schwassmann-Wachmann 1       13 ?     40 N to 70 S (best morning, 02:30)
•  104P/Kowal 2                     13.5     50 N to 15 S (early evening)
•  78P/Gehrels 2                    14 ?     65 N to 35 S (evening)
•  43P/Wolf-Harrington              14 ?     25 N to 65 S (best 22:00)
```

Figure 4.12. BAA Comet Section Homepage.

Figure 4.13. The BAA observations for Comet Hartley2.

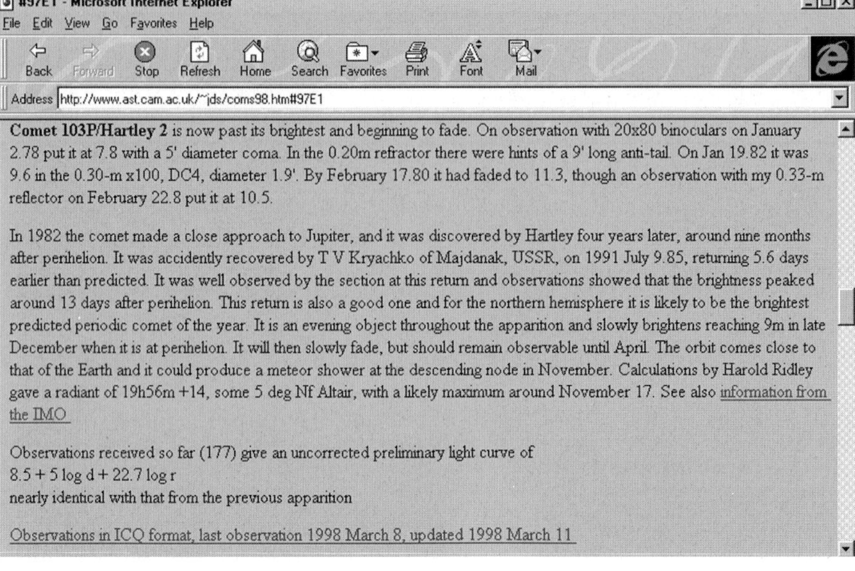
#97E1 - Microsoft Internet Explorer

File Edit View Go Favorites Help

Back Forward Stop Refresh Home Search Favorites Print Font Mail

Address http://www.ast.cam.ac.uk/~jds/coms98.htm#97E1

Comet 103P/Hartley 2 is now past its brightest and beginning to fade. On observation with 20x80 binoculars on January 2.78 put it at 7.8 with a 5' diameter coma. In the 0.20m refractor there were hints of a 9' long anti-tail. On Jan 19.82 it was 9.6 in the 0.30-m x100, DC4, diameter 1.9'. By February 17.80 it had faded to 11.3, though an observation with my 0.33-m reflector on February 22.8 put it at 10.5.

In 1982 the comet made a close approach to Jupiter, and it was discovered by Hartley four years later, around nine months after perihelion. It was accidently recovered by T V Kryachko of Majdanak, USSR, on 1991 July 9.85, returning 5.6 days earlier than predicted. It was well observed by the section at this return and observations showed that the brightness peaked around 13 days after perihelion. This return is also a good one and for the northern hemisphere it is likely to be the brightest predicted periodic comet of the year. It is an evening object throughout the apparition and slowly brightens reaching 9m in late December when it is at perihelion. It will then slowly fade, but should remain observable until April. The orbit comes close to that of the Earth and it could produce a meteor shower at the descending node in November. Calculations by Harold Ridley gave a radiant of 19h56m +14, some 5 deg Nf Altair, with a likely maximum around November 17. See also information from the IMO

Observations received so far (177) give an uncorrected preliminary light curve of
8.5 + 5 log d + 22.7 log r
nearly identical with that from the previous apparition

Observations in ICQ format, last observation 1998 March 8, updated 1998 March 11

The ephemeride tables enable you to find the latest cometary data for loading into your own sky planetarium program, allowing you to print your own finder charts and plan your observations. Not the most "gee-whizz" of web sites with nothing to take your breath away. Just useful cometary data, as you would expect from such a traditional source as the BAA, but not to be criticised for it.

Cometwatch

Our final section in this chapter is one I have on my "favourites" list. If you have ever looked at the photo/image credits in magazines for the first image of a new comet, whose name will you see more than any other? The answer, I bet, would be USA amateur Tim Puckett. Tim runs the Cometwatch program with its own dedicated Homepage. It is not difficult to see why this talented comet imager is usually first on the scene. Tim has built his own observatory with not one but three telescopes. All are optimised on comet imaging and

Figure 4.14. The Cometwatch Homepage – background image Comet Tempel-Tuttle.

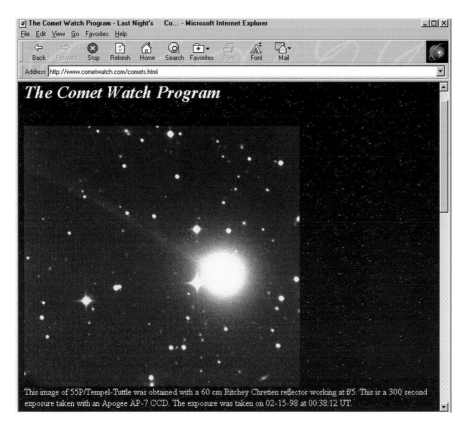

monitoring. When the third telescope comes on line Tim's chances of discovering one himself will be dramatically increased. This telescope, or more accurately camera, is a giant Baker-Nunn Schmidt camera with a 20 inch aperture and 31 inch primary mirror operating at f/1 – yes, a focal ratio of one!

To be able to locate, lock onto and track comets, all Tim's telescopes are computer controlled. Tim uses Comsoft's professional Telescope Control System, PC-TCS, in conjunction with TPoint, (see Chapter 9 for details of both). This allows unguided imaging of comets no matter what their movement across the sky. Also, because much of the operation can be automated, a whole series of images can be taken in rapid succession, often as many as 100, without much difficulty. These can then be assembled to make a movie, showing the movement of the comet in front of the background stars. These movies have brought Tim much fame, having been shown on the USA TV show, Good Morning America. A selection of Tim's images and animations are included on the CD-ROM.

If it is the latest comet image you are after, Cometwatch is probably the place to look first – there might even be a video of it too!

Figure 4.15. Listings of available comet images.

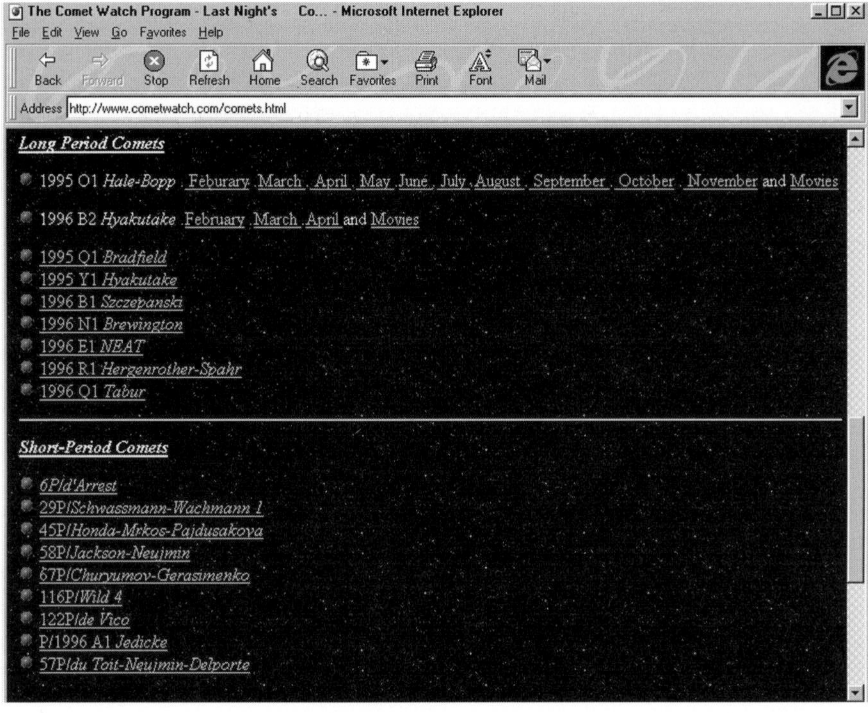

URLs Featured in this Chapter

JPL Solar Systems Dynamics Group (HORIZONS)
http://ssd.jpl.nasa.gov

Views of the Solar System
http://bang.lanl.gov/solarsys

Mooncalc by Dr. Monzur Ahmed
http://www.starlight.demon.co.uk/mooncalc

Minor Planet Center (MPC)
http://cfa-www.harvard.edu/cfa/ps/mpc.html

BAA Comet Section
http://www.ast.cam.ac.uk/~jds

Cometwatch Homepage, Tim Puckett
http://www.cometwatch.com

Other URLs not Referred to but Recommended

Asteroid Data Service by Lowell Observatory
http://asteroid.lowell.edu/
Huge asteroid database

Fred Espenak's 12 Year Planetary Ephemerides
http://planets.gsfc.nasa.gov/TYPE/TYPE.html

Fred also does a 100 year Moon Phase page
http://planets.gsfc.nasa.gov/phase/phasecat.html

Charles S. Morris's Comet Observation Homepage
http://enke.jpl.nasa.gov/

Gary Kronk's Comet Page
http://medicine.wustl.edu/~kronkg/comet.html

Chapter 5 Data Sources III – Eclipses and Occultations

Background

At the time of writing, interest in the 1999 Total Eclipse of the Sun, visible from the extreme south-west of England and across Europe, is already reaching fever pitch. Hotel bookings along the path of totality are virtually all gone but there will be another way to watch this (and subsequent ones for that matter). What I am talking about is web broadcasts. Due to a technique called streaming you can have images, even sound and movies (with an appropriate software plug-in), live within your web-browser. Video cameras broadcasting to the Internet tend to be known as "Webcams". Here in the UK Michael Oates, who runs the excellent web site for the society of Popular Astronomy, was probably the first to show live images of the 1996 partial eclipse. The following year, the 1997 total eclipse took place and there was the first successful attempt to capture a total eclipse live and broadcast it on the Internet. This service was organised by a Japanese group called LIVE!ECLIPSE and as the eclipse took place in Mongolia and Siberia, it brought a remote event to many who otherwise could not have seen it. From now on you can expect every eclipse, including the 1999 one, to spawn numerous web sites covering them live online.

Despite the hype over eclipses these are relatively rare events but there are more frequent events of a similar, if less spectacular, nature. I am referring to occultations. Strictly speaking eclipses are just a special form of occultation. Of these, the most common are the lunar ones and there is an International Occultation Timing Association Centre (IOTA) which collects and analyses data. This chapter includes a look at their web site and what services they offer. Finally we try some excellent software for lunar occultation predictions.

NASA/Goddard Space Flight Center – Eclipse Homepage

This really is the definitive site for Eclipses and anyone a little bit interested would do well to bookmark it. Run by Fred Espenak for the NASA/Goddard Space Flight Center, it covers everything you could possibly wish to know. The "eclipse latest" section covers imminent eclipses (solar and lunar). Publications are available for every eclipse indivi-

Figure 5.1. Fred Espenak's Eclipse Homepage.

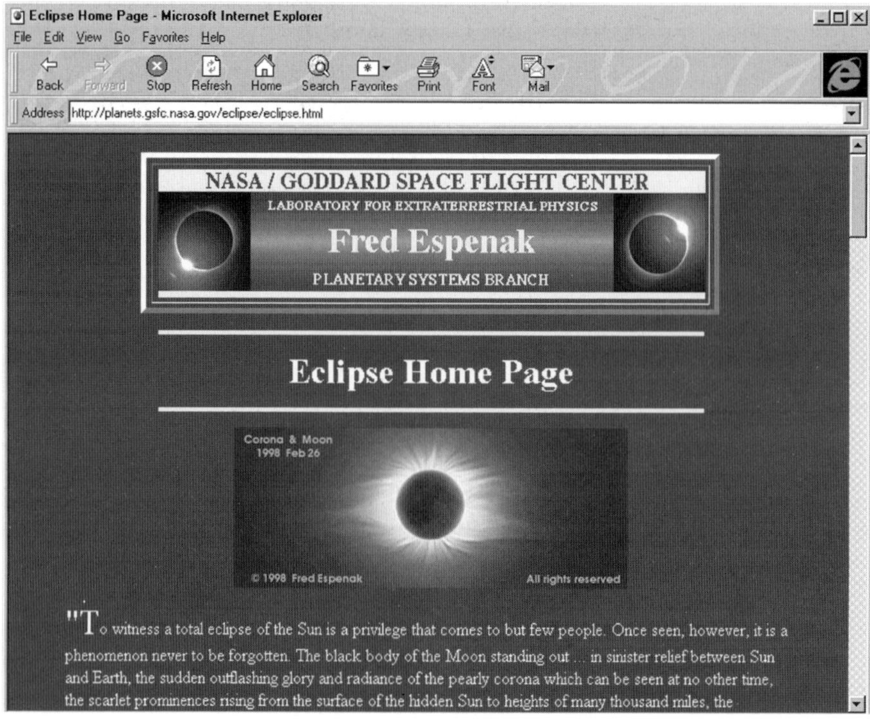

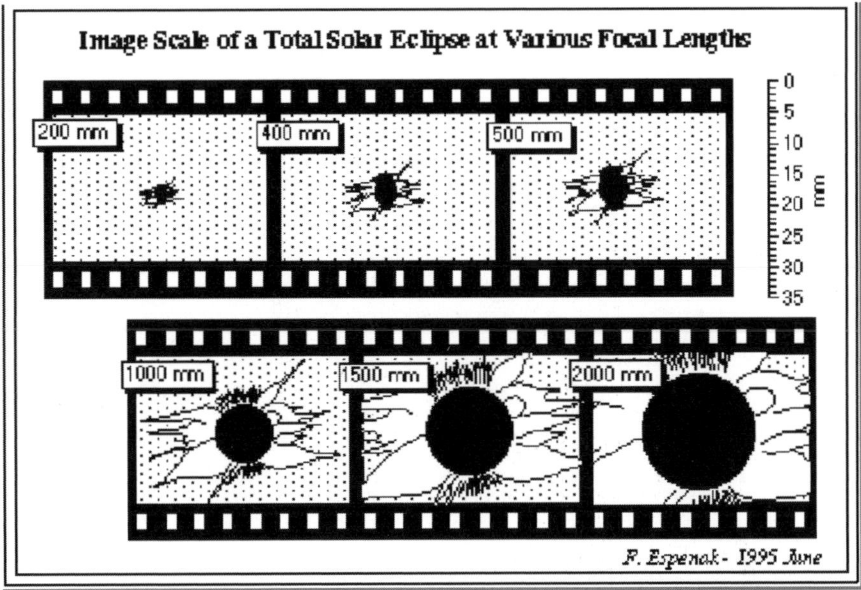

Figure 5.2. Fred Espenak's drawing showing the size of the solar image for various lens focal lengths.

Included on CD-ROM

dually plus handbooks for each year. Of particular interest, in view of our own forthcoming eclipse, is an excellent guide to safe observing. For those wishing to record the event for posterity, there are tips both for photographers and those with video cameras. You only get one chance with eclipse photography so these guides are essential reading. It will be a long time before you can retake any pictures if your exposure is wrong! Figures 5.2 and 5.3 (*overleaf*) are extracts from the astrophotography guide. The guide is included on the CD-ROM

An often overlooked but critical aspect of any eclipse is weather prediction. Imagine travelling around the world only to be clouded out! Again Fred comes to our rescue producing tables with the average number of clear days for sites on the eclipse path, for that particular time of year. Not a guarantee of success but a practical way of maximising the observing opportunities and minimising the risks of what could only be described, if you can forgive the pun, as a total disaster.

If you are hooked on eclipses and want to plan your future holidays around them then the site publishes tables and, probably more importantly, maps for all forthcoming eclipses for up to the year 2020. These show precisely when and where they will take place for

TABLE 2

SOLAR ECLIPSE EXPOSURE GUIDE

ISO		f/Number						
25		1.4	2	2.8	4	5.6	8	
50		2	2.8	4	5.6	8	11	
100		2.8	4	5.6	8	11	16	
200		4	5.6	8	11	16	22	
400		5.6	8	11	16	22	32	
800		8	11	16	22	32	44	
1600		11	16	22	32	44	64	
Subject	**Q**	**Shutter Speed**						
Solar Eclipse								
Partial[1] -4.0 ND	11	—	—	—	1/4000	1/2000	1/1000	1
Partial[1] -5.0 ND	8	1/4000	1/2000	1/1000	1/500	1/250	1/125	
Baily's Beads[2]	11	—	—	—	1/4000	1/2000	1/1000	1
Chromosphere	10	—	—	1/4000	1/2000	1/1000	1/500	1

Figure 5.3. A small extract from Fred Espenak's exposure guide for solar eclipse photography. The full version is on the CD-ROM.

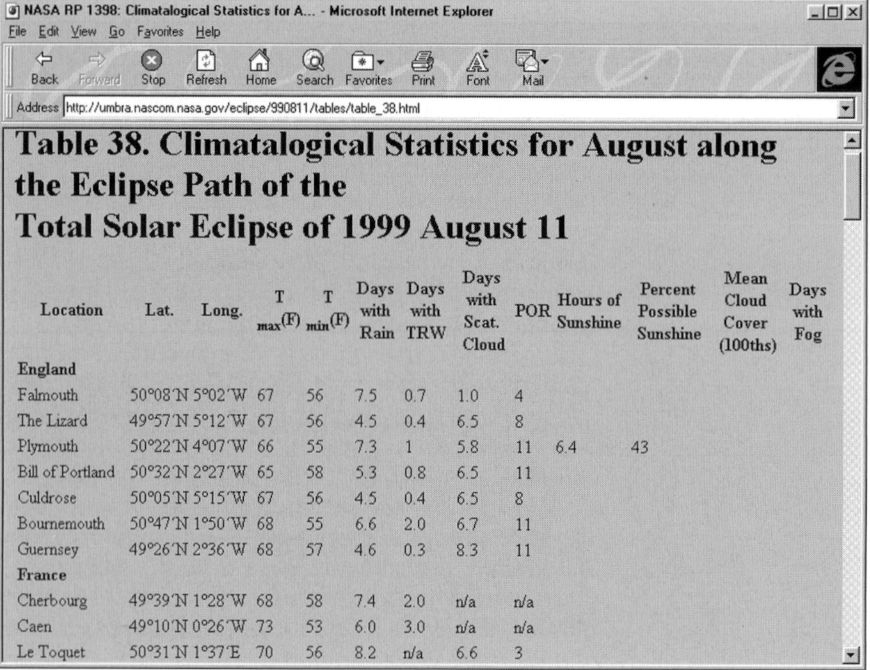

Table 38. Climatalogical Statistics for August along the Eclipse Path of the Total Solar Eclipse of 1999 August 11

Location	Lat.	Long.	T max(F)	T min(F)	Days with Rain	Days with TRW	Days with Scat. Cloud	POR	Hours of Sunshine	Percent Possible Sunshine	Mean Cloud Cover (100ths)	Days with Fog
England												
Falmouth	50°08'N	5°02'W	67	56	7.5	0.7	1.0	4				
The Lizard	49°57'N	5°12'W	67	56	4.5	0.4	6.5	8				
Plymouth	50°22'N	4°07'W	66	55	7.3	1	5.8	11	6.4	43		
Bill of Portland	50°32'N	2°27'W	65	58	5.3	0.8	6.5	11				
Culdrose	50°05'N	5°15'W	67	56	4.5	0.4	6.5	8				
Bournemouth	50°47'N	1°50'W	68	55	6.6	2.0	6.7	11				
Guernsey	49°26'N	2°36'W	68	57	4.6	0.3	8.3	11				
France												
Cherbourg	49°39'N	1°28'W	68	58	7.4	2.0	n/a	n/a				
Caen	49°10'N	0°26'W	73	53	6.0	3.0	n/a	n/a				
Le Toquet	50°31'N	1°37'E	70	56	8.2	n/a	6.6	3				

both solar and lunar eclipses. Figure 5.5 (*overleaf*) is the world map of the path for the solar eclipses of 2001. Shown at this scale it hardly does the map justice but at least it gives an indication of the slick presentation of data from this site. These guides for eclipses up to the year 2020 are included on the CR-ROM, and at actual published size, so you will have no difficulty reading them and getting your plans made early.

Included

on CD-ROM

As well as eclipse information online, NASA produces printed guides for each total eclipse. The guide to the total solar eclipse of August 1999 written by Fred Espenak and Jay Anderson runs to 128 pages. It is crammed full of predictions, timings, maps plus of course where the best weather is likely to be! In keeping with the USA policy of dissemination of information it is free (postage only payable). If only we had similar foresight here in the UK. However, because public demand for it was so huge they were out of print almost as soon as it was published. Fortunately the Internet came to the rescue. As well as being published conventionally, they are also published online in both HTML and Acrobat format. The latter preserves page formatting so it is a perfect facsimile of the printed version. They are available from NASA's ftp (file transfer protocol) site but modern browsers will be able to download them directly. The URL for this service is given at the end of this chapter. The only snag is that as the guides are so long, file sizes can be over 18 Mb.. They have free local phone calls in the USA so, although downloading could take 3 hours or more, telephone costs would not preclude it. Either way it is not a problem for owners of this book as the full Acrobat version of the guide is included on the CD-ROM. Eclipse watchers owe Fred Espenak a big vote of thanks for this and his excellent site.

Included

on CD-ROM

◄ **Figure 5.4.** Extract from the table of weather statistics for the 1999 eclipse. Perhaps England will not be the best place to see it!

LIVE!ECLIPSE

As mentioned in the introduction to this chapter, a group of Japanese enthusiasts were probably the first to broadcast a total eclipse live on the Internet. Thousands of students in Japan were able to watch an event otherwise impossible for them to see. They also covered the 1998 eclipse and Figure 5.6 (*overleaf*) was downloaded just after the event. At the time of writing

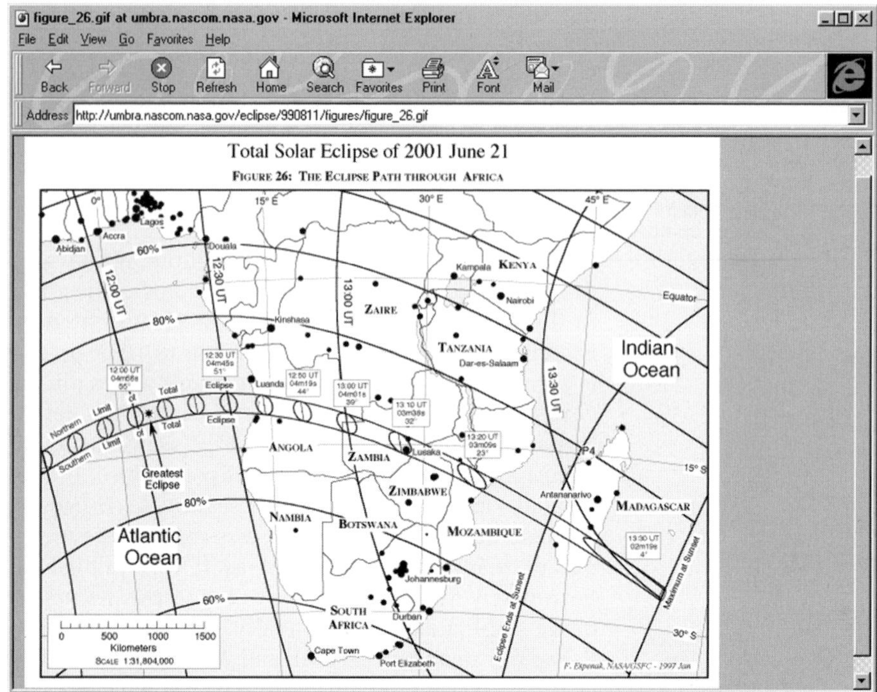

Figure 5.5. Map of Africa showing the solar eclipse for the year 2001.

their plans for the 1999 one have yet to be announced. No doubt they will be there and, if past form is a guide with their successful previous broadcasts, they will definitely be worth keeping a close eye upon as the big day draws near. So if you cannot get to the 1999 eclipse, or are clouded out, they should provide a viable alternative. Figure 5.7 (*overleaf*) is just a sample of the many fine images this group produced in 1998.

Society for Popular Astronomy

Eclipses live on the Internet

Although the Japanese LIVE!ECLIPSE group were the first for a total eclipse, they were actually not the first for any type of eclipse. They were beaten to it by Michael Oates, who runs the web site for the Society of Popular Astronomy (SPA), for the 1996 partial eclipse visible from Britain. Michael hooked a telephoto lens up to his video camera and produced images for

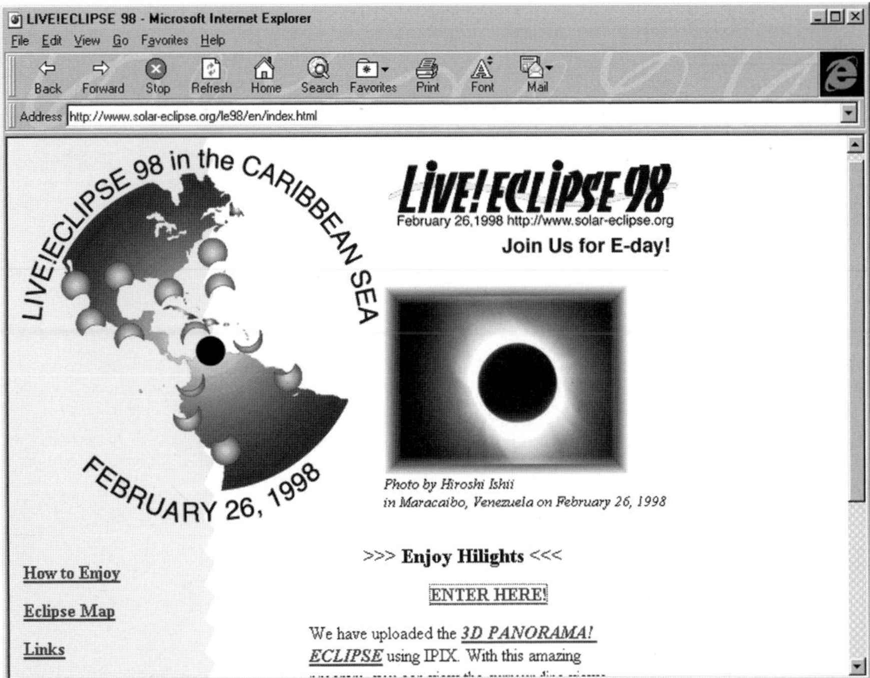

Figure 5.6. LIVE!ECLIPSE Homepage.

display on the web pages as the event unfolded. He had also enlisted a team of helpers (CCD imagers Maurice Gavin, David Strange and John Jones). This was fortunate as Michael himself was clouded out just before maximum, so he was unable to provide images himself for the full duration. During the event nearly 5000 people tried to access the site – tried being the operative word! The site was swamped with traffic, a victim of its own success. No doubt the 1999 eclipse will be covered by Michael for the SPA, so if the clouds come then log on to the SPA site – rumour has it this time the server will be up to the strain!

International Occultation Timing Association (IOTA)

The timing of cccultations is one of the few remaining areas in Astronomy where the amateur can still play an important part and you are in fact actively encouraged to get involved. The International Occultation Timing

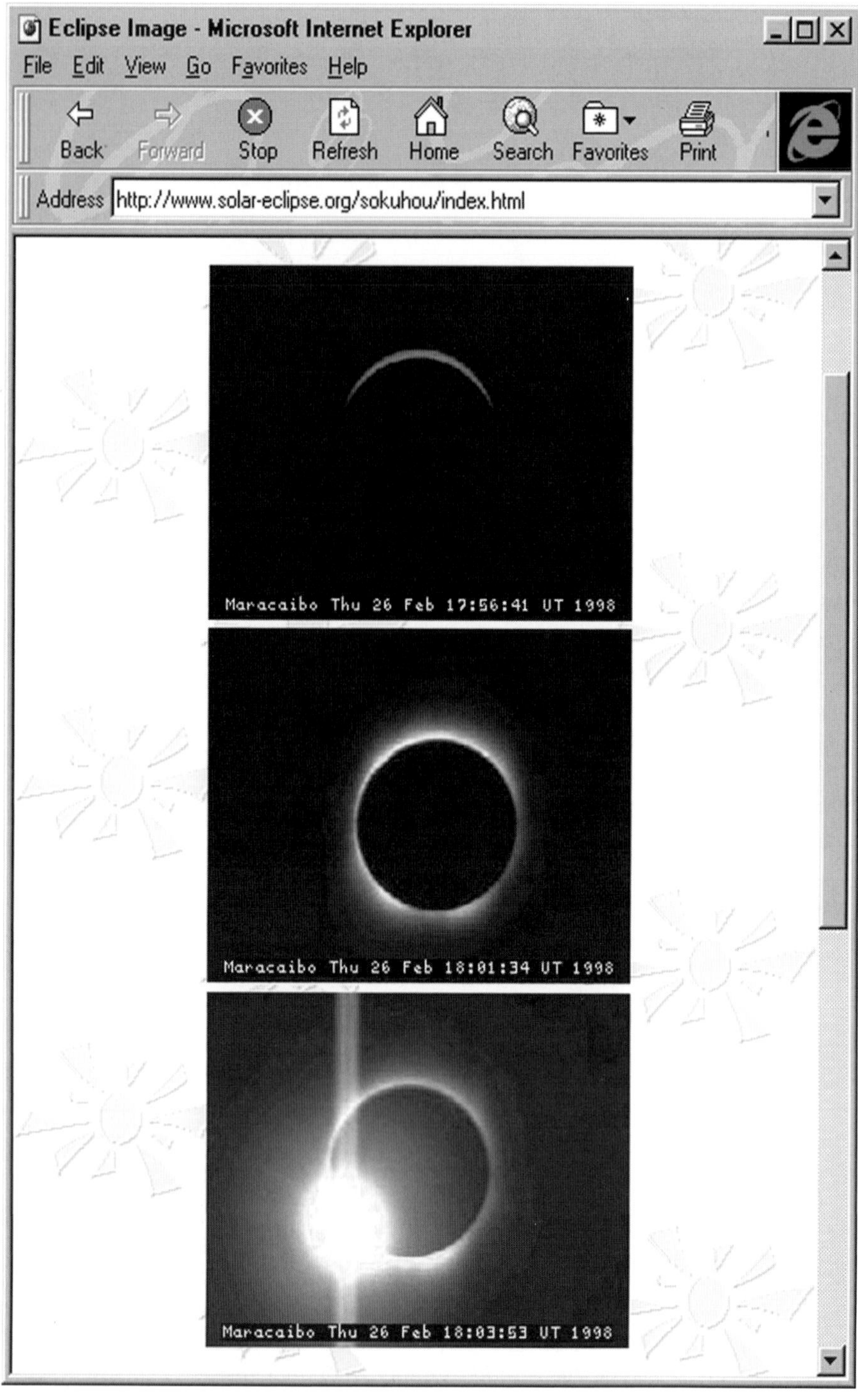

Figure 5.7. A sample of the fine images of the 1998 eclipse taken by the LIVE!ECLIPSE group.

The Society for Popular Astronomy - Microsoft Internet Explorer

Address http://www.u-net.com/ph/spa/home.htm

The Society for Popular Astronomy

Starstruck?

Then the **Society for Popular Astronomy** is for you! All the fun of astronomy at a price you can afford!

- What's New NEW
- Popular Astronomy magazine
- News Circulars
- SPA Meetings and Events
- Observing Sections
- Activities
- SPA Officers
- Help and Advice
- Book Scheme

Partial Eclipse of the Sun - Microsoft Internet Explorer

Address http://www.u-net.com/ph/spa/eclipse/pr-002.htm

Society for Popular Astronomy

LIVE ECLIPSE ON THE INTERNET

AN ECLIPSE of the Sun over the UK next week - the biggest from this country since 1961 - **will be the first ever to be broadcast live on the Internet**.

The Society for Popular Astronomy, Britain's leading society for space enthusiasts, has set up a page on the World Wide Web with all you need to know to observe the spectacle on Saturday 12th October.

And in a unique operation, cameras around the country will feed regularly updated images of the eclipse to the page. If the sky is clear from at least one of them, astronomers from anywhere in the world will be able to log on to view the event.

The Society for Popular Astronomy's eclipse page can be found at: http://www.u-net.com/ph/spa/eclipse/partial.htm

For the UK, this eclipse is the prelude to the Big One - the total eclipse due on 11 August 1999.

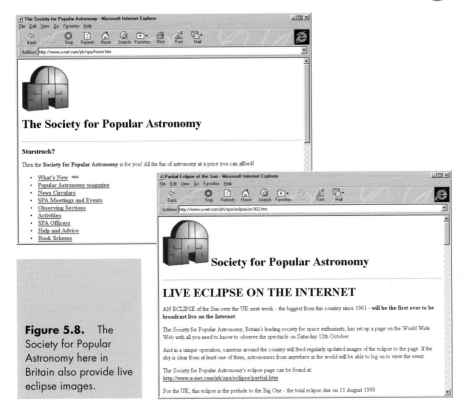

Figure 5.8. The Society for Popular Astronomy here in Britain also provide live eclipse images.

Centre (IOTA) is the body which oversees occultations and encourages their observation and recording. Lunar occultations, particularly grazing ones, are of significant importance as they can provide an accurate plot of the Moon's profile. If you are as old as me, you will recall that it was occultations by the Moon which enabled the position of the first quasar to be accurately determined.

In the case of asteroid occultations, also covered by IOTA, they can reveal much about an asteroid which would be impossible to discover by virtually any other means – short of sending a space probe to it. In the past teams of observers located on different eclipse paths have, by accurate timing of asteroids occulting a background star, been able not only to work out the asteroid's size but its shape too. In a few cases where there has been reappearance of the star part way through the eclipse, this provides strong evidence for the asteroid being two objects not one.

IOTA is perhaps not, at least at the time of writing, totally "switched on" with regard to the Internet but their homepage (Figure 5.9, *overleaf*) does provide a point of contact and some introductory information.

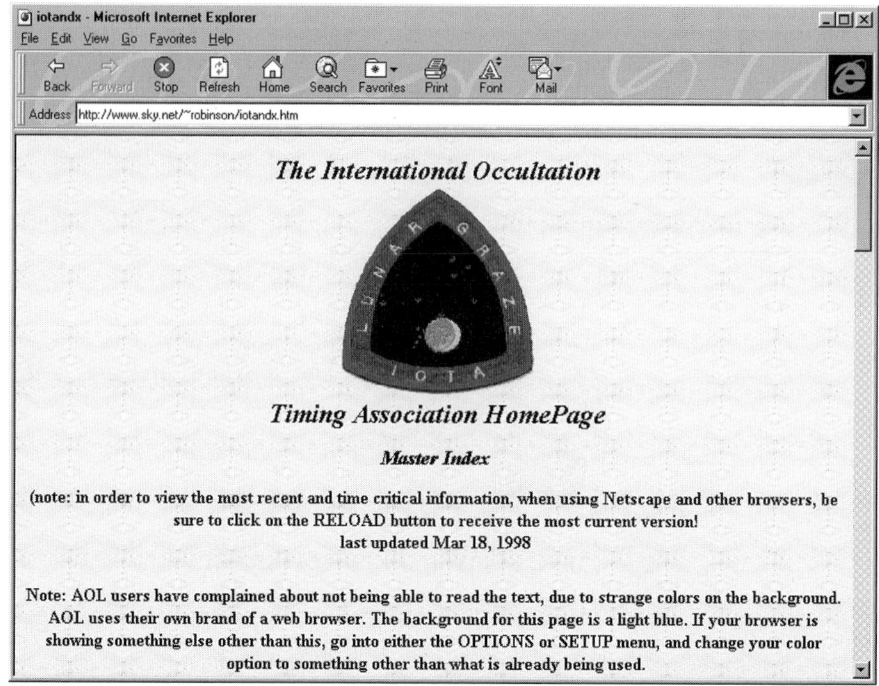

Figure 5.9. The IOTA Homepage.

To obtain their newsletter and their detailed finder charts you have to subscribe. However they do produce a guide to Grazing Occultations which, together with their introduction to themselves, is included on the CD-ROM. An example of one of their lunar grazing plots is shown in Figure 5.10 (*opposite*).

Lunar Occultation Workbench Software

Lunar Occultation Workbench, or LOW as it is known, is a super freeware program for predicting occultations of the planets and stars by the Moon. Its stellar database comprises the 53,879 stars in the XZ94D catalogue, which has stars down to magnitude 10. Written by Eric Limburg of the Dutch Occultation Association, it graphically represents the Moon and star/planet at the moment of disappearance and/or re-appearance. The position angle of the disappearance/reappearance is shown as is the phase of the Moon (Figure 5.11, *overleaf*). Other relevant data is given in tabular form.

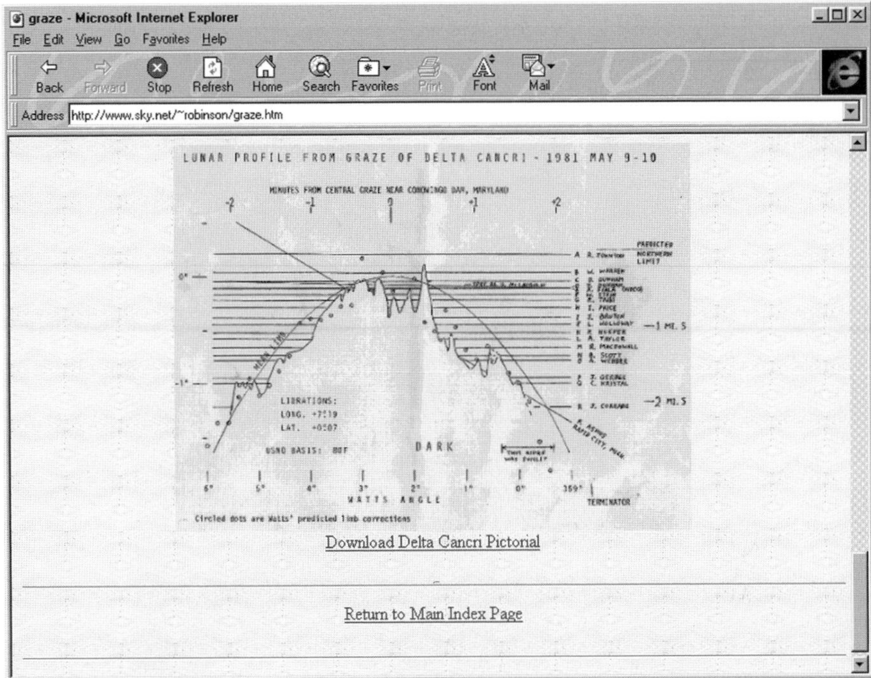

Figure 5.10. Extract from *IOTA Grazing Occultations Guide*. Lunar profile from observing the occultation of the star Delta Cancri.

The first time you run the program, details of your observing site and equipment have to be entered. This takes some time but, if the data is saved, it will not need to be re-entered the next time you run it. The program includes a "wizard" to guide beginners through the set-up procedure and I would recommend you make use of it. Once set-up it calculates all occultations for your site for a full year. It is then a simple task to "tab" through the list and, when you find one of interest, display details together with a graphical representation of the event.

The standard version of the program (version 1.3) is included on the CD-ROM and is ideal for general occultation viewing. It requires Windows 95/98/NT and about 20 Mb of disc space. An even larger version of the software, called "complete", is also included on the CD-ROM but is 30 Mb in size. It is optimised for grazing occultations which require a higher accuracy and has, as a result, longer calculation times. The standard version is probably more appropriate for the average amateur but, whichever version you try, I am sure you will be impressed with the facilities and results. In both cases, our thanks must go to Eric Limburg for making these excellent programs available to us as freeware.

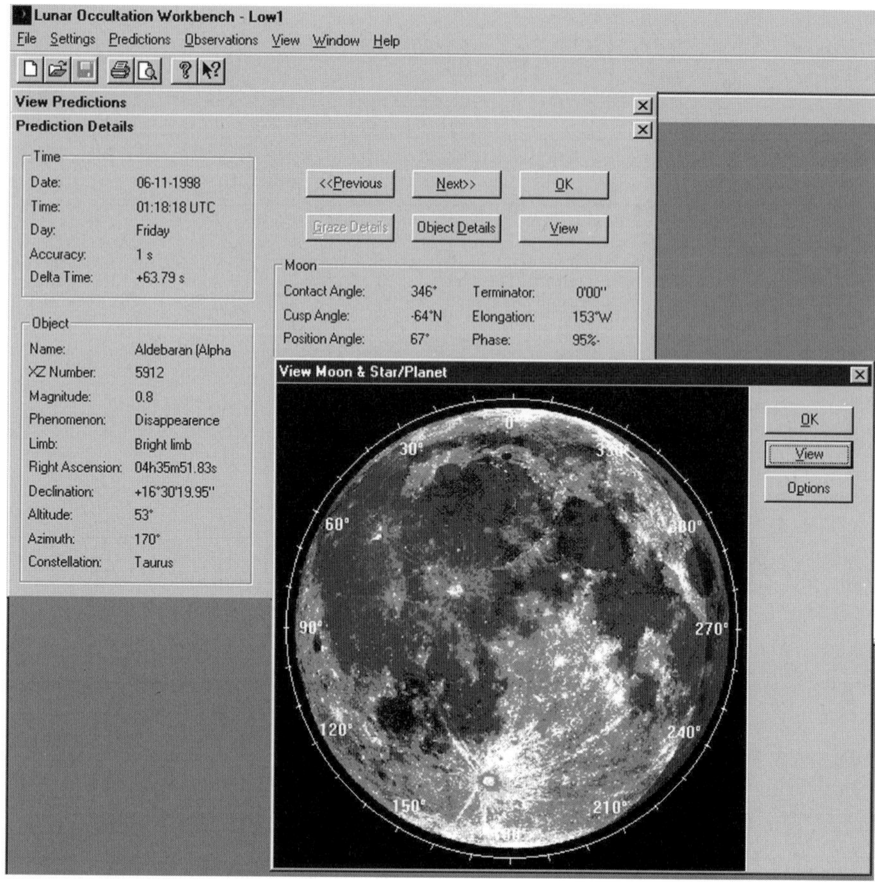

Figure 5.11. Lunar Occultation Workbench software – prediction of the occultation of Aldebaran, 6th November 1998.

URLs Featured in this Chapter

NASA/Goddard Space Flight Center
Fred Espenak's Eclipse Homepage
http://planets.gsfc.nasa.gov/eclipse

Ditto but their FTP site for downloads
ftp://umbra.nascom.nasa.gov/eclipse

LIVE!ECLIPSE Homepage
http://www.solar-eclipse.org

Society for Popular Astronomy (SPA)
Michael Oates' Eclipse Pages
http://www.u-net.com/ph/spa/home.htm

International Occultation Timing Association (IOTA)
http://www.sky.net/~robinson/iotandx.htm

Dutch Occultation Association – LOW software
http://web.inter.nl.net/hcc/elimburg.doa/software.htm

Other URLs not Referred to but Recommended

Sky & Telescope Occultation Page
(Despite the URL address nothing to with the Occult!)
http://www.skypub.com/occults/occults.html

Europe Astronomy Online – live eclipse broadcasts
http://www.eso.org/astronomyonline

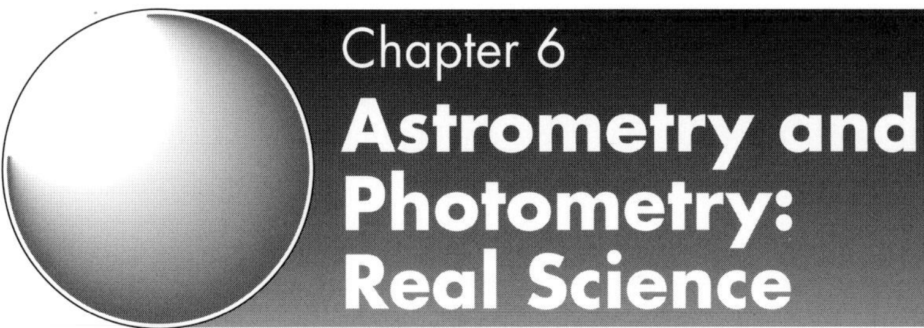

Chapter 6 Astrometry and Photometry: Real Science

Background

Astrometry, the measurement of position, and Photometry, the measurement of brightness, have undergone a dramatic increase in interest. Two events have probably been responsible for this. Firstly the European Space Authority's Hipparcos Space Astrometry Mission, which produced a celestial treasure trove of astrometric and photometric data, the accuracy of which had never been seen before. The second was the advent of affordable CCD cameras for the amateur. This meant for the first time astrometry and photometry were no longer the preserve of the specialist with exotic equipment. The ordinary amateur could now do real science with his backyard telescope and CCD camera. These devices provide ready digitised images from which, using appropriate software, position and brightness can be measured.

This Chapter can unfortunately only touch the surface, in fact the Hipparcos Mission and its results could justify a book on its own. However, such is the significance of Hipparcos that I have devoted the majority of the chapter to it. It is of course fundamental to all aspects of astronomy to have accurate positions and distances to stars. They form the building blocks on which virtually all other research is based. Already accepted distances to well known objects have had to be revised in the light of Hipparcos. Hipparcos is still very new (the catalogues were completed August 1996 and published the following year) and no doubt over the next few years we will see much imaginative use and display of this data. I can foresee the day when accurate 3-D space flight simulations of a journey to a star of your choice are available. There are exciting times ahead.

The Hipparcos Space Astrometry Mission (ESA)

Launched by the European Space Authority in August 1989, the Hipparcos mission was dedicated to the precise measurement of the positions, parallaxes and proper motions of stars. Only a satellite above the distortions of our atmosphere could achieve the accuracy required. Its work took over three years and was successfully completed by August 1993. Its primary task was to measure five astrometric parameters for 120,000 stars to a precision of 2 to 4 milliarcseconds. Its other task, the Tycho Experiment, was to measure the astrometric (to a lower precision) and two-colour

Figure 6.1. The Hipparcos Homepage – an excellent example of how to make complex scientific data accessible and understandable to the public.

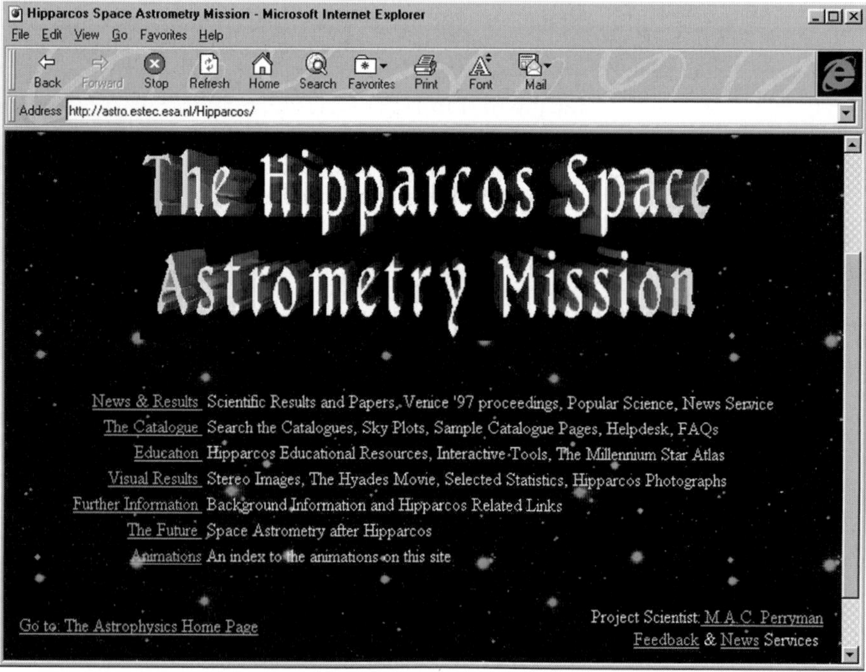

photometric properties of another 400,000 stars. It comfortably exceeded its targets despite a series of problems. The outcome was the Hipparcos Catalogue of 120,000 stars with 1 milliarcsecond positional accuracy and the Tycho Catalogue, with over 1 million stars to 20 to 30 milliarcsecond accuracy and with two colour photometry.

Now the good news, the Hipparcos Catalogues are now freely available. Now the bad news, with a size of around one Terabyte (one trillion bytes), these catalogues are a little on the large size and will not fit on our CD-ROM (only highlights have been included). Fortunately there is an alternative through the excellent Hipparcos Homepage (Figure 6.1). It is an outstanding example of how to make complex scientific data accessible and understandable to the general public. Unfortunately, it has become a victim of its own success, with the site (at least at the time of writing) groaning under the weight of people accessing it. To speed you in accessing this site, the Introduction and Frequently Asked Questions pages have already been downloaded for you and are to be found on the accompanying CD-ROM. They make excellent background reading.

Included on CD-ROM

In addition, the Hipparcos team have published selected information in what is known as "Table 3.6". This includes the results for the 150 stars closest to the Sun, the 150 stars with the largest proper motions, the 150 stars with the largest transverse velocity and 150 stars with the highest absolute luminosity. Note the distinction between proper motion and actual transverse velocity. The latter is, of course, a function of a star's distance as well as its apparent proper motion. Only those stars which could be measured with the necessary precision have been included. This Table 3.6 has been downloaded and is included on the CD-ROM. It is an excellent reference source and included in it is an explanation of the fields. Our thanks to Michael Perryman and the Hipparcos team for agreeing to this.

Included on CD-ROM

If the full Catalogues are too big, how do we get at the data? Again this excellent homepage provides us with some tools, namely Sky Plot, which is a graphical, easy to use Java application.

This is undoubtedly the best way to get at the data although, as we will see later, there is an alternative. Sky Plot is a two stage affair, first we must select an area of sky to display the catalogued stars from Hipparcos and/or Tycho, then we can interrogate them

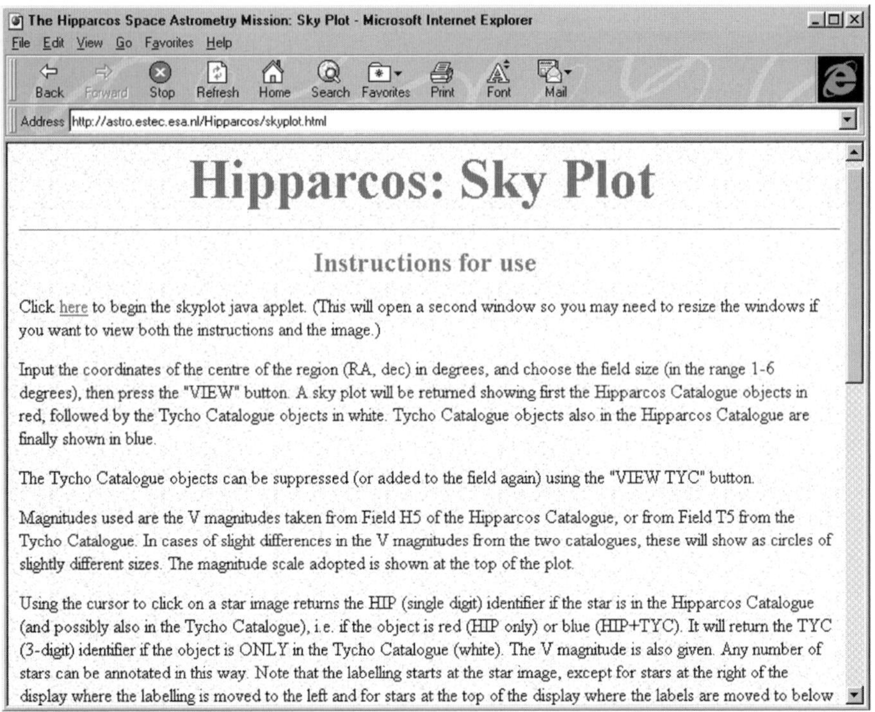

Figure 6.2. Hipparcos Sky Plot data access system.

individually to find out their properties. Figure 6.3 (*opposite*) shows just such a plot for the Hyades cluster in Taurus. Aldebaran is the bright star to left and the V of the cluster is easily discernible. To select an area of sky we have to know the RA and DEC co-ordinates of the area in question. Here there is a little quirk, RA is in degrees not hours/minutes as is usual. The area to be displayed has also to be selected from a pull down list. Six degrees is the biggest but here a word of warning. If accessing the site at popular times, loading stars for a six by six degree field will try your patience. Either select a smaller field or log-on at 7:00 am! Figure 6.3 shows the field displayed, although not evident in the black and white reproduction here, colour coding is used to show whether a star has come from the Hipparcos and/or Tycho catalogues.

Having displayed the stars, to get information on a particular one, simply click on the star with the mouse and press the "get information" button. It is that easy. Figure 6.4 (*overleaf*) is the result of clicking on Aldebaran, only the first 18 fields are shown here. Aldebaran is of course not a member of the Hyades but a foreground object. You will note that Hipparcos uses

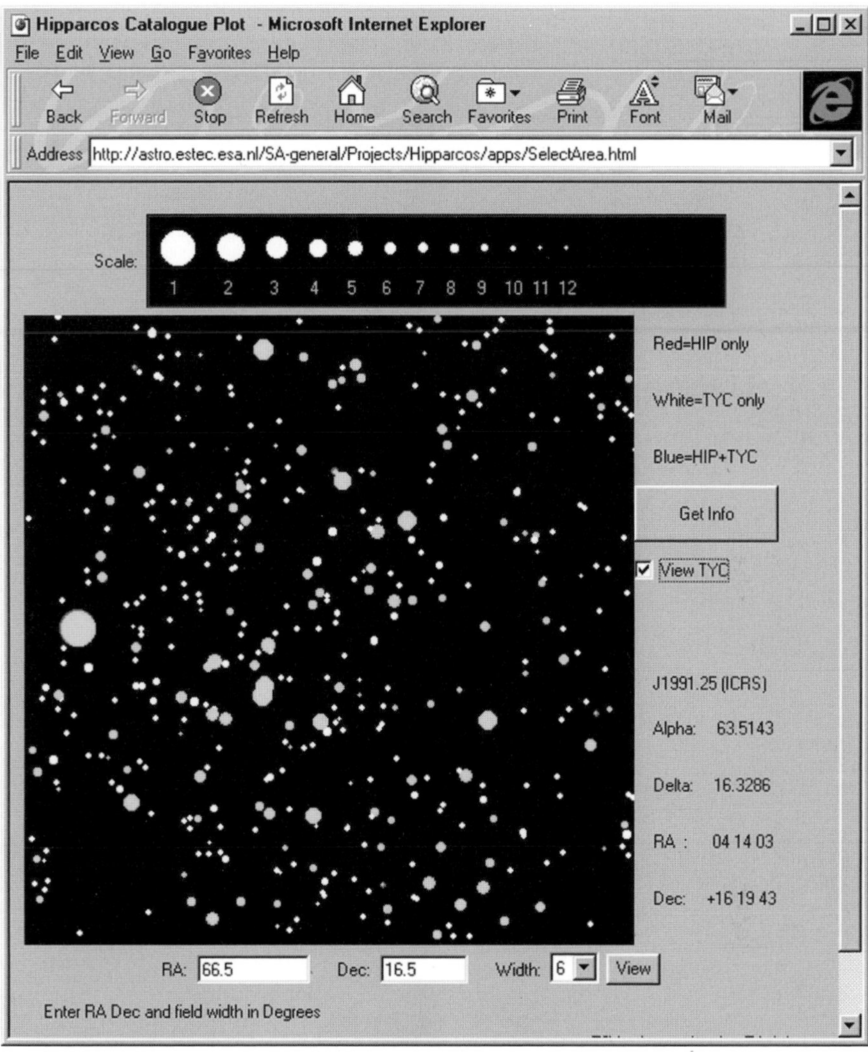

Figure 6.3. Sky Plot and the Hyades star cluster.

its own star numbering system, known as the HIP number, so Aldebaran is catalogued as HIP21421. Its magnitude is given as 0.87.

As mentioned earlier, there is an alternative way of accessing the Hipparcos data. One such facility is VizieR, a service provided by the Centre de Donnees astronomiques de Strasbourg (CDS). This is not limited to Hipparcos, being in fact a general purpose service for interrogating most published astronomical catalogues. This has one advantage over Sky Plot in that individual stars can be interrogated directly, rather than the slower process of displaying an area

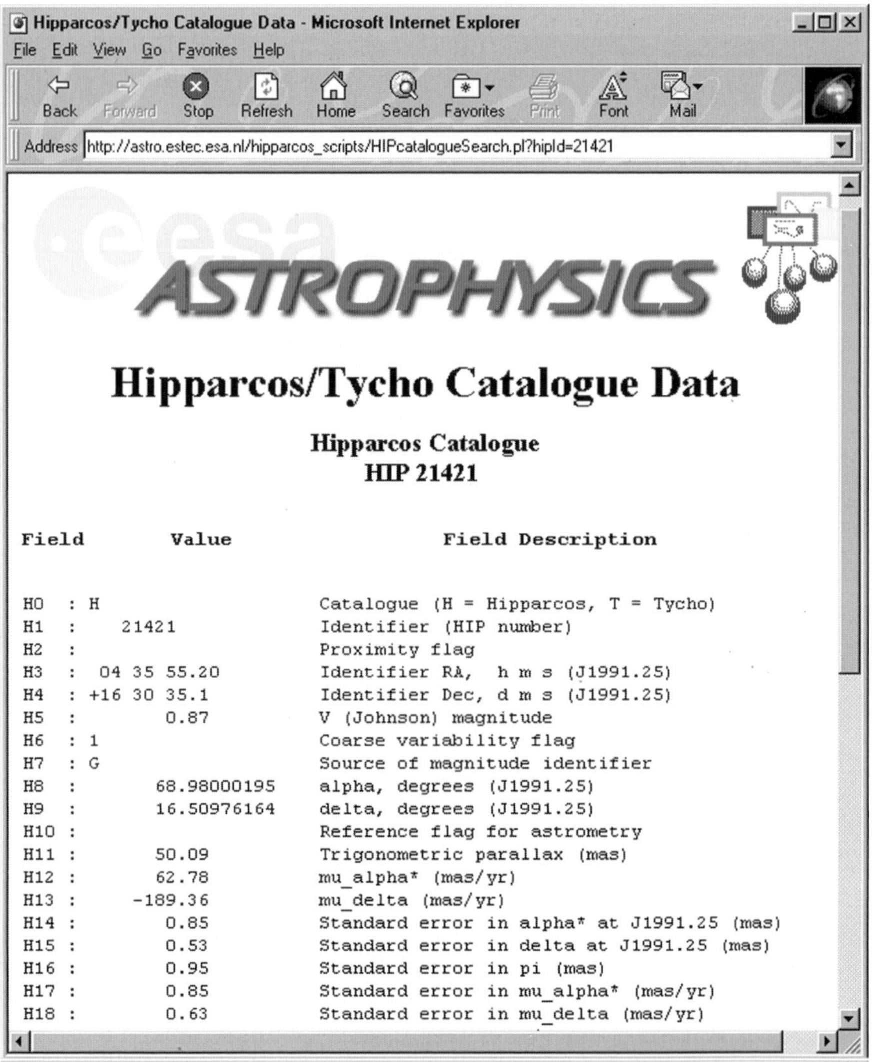

Hipparcos/Tycho Catalogue Data - Microsoft Internet Explorer

Address http://astro.estec.esa.nl/hipparcos_scripts/HIPcatalogueSearch.pl?hipId=21421

ASTROPHYSICS

Hipparcos/Tycho Catalogue Data

Hipparcos Catalogue
HIP 21421

Field	Value	Field Description
H0 :	H	Catalogue (H = Hipparcos, T = Tycho)
H1 :	21421	Identifier (HIP number)
H2 :		Proximity flag
H3 :	04 35 55.20	Identifier RA, h m s (J1991.25)
H4 :	+16 30 35.1	Identifier Dec, d m s (J1991.25)
H5 :	0.87	V (Johnson) magnitude
H6 :	1	Coarse variability flag
H7 :	G	Source of magnitude identifier
H8 :	68.98000195	alpha, degrees (J1991.25)
H9 :	16.50976164	delta, degrees (J1991.25)
H10 :		Reference flag for astrometry
H11 :	50.09	Trigonometric parallax (mas)
H12 :	62.78	mu_alpha* (mas/yr)
H13 :	-189.36	mu_delta (mas/yr)
H14 :	0.85	Standard error in alpha* at J1991.25 (mas)
H15 :	0.53	Standard error in delta at J1991.25 (mas)
H16 :	0.95	Standard error in pi (mas)
H17 :	0.85	Standard error in mu_alpha* (mas/yr)
H18 :	0.63	Standard error in mu_delta (mas/yr)

Figure 6.4. An extract from the Hipparcos/Tycho database for Aldebaran.

then querying on an individual one. You will of course need to know the HIP number of the star in question but even here CDS can help (try SIMBAD) with its extensive cross-reference facilities. The URL for CDS and VisieR are included at the end of this chapter.

We return again to the Hipparcos web site proper, where there are even more goodies and facilities on offer. We will have a look at a few more of them. Firstly, High Proper Motion Stars. Hipparcos measured the proper motion of its target stars, this being the apparent angular movement per year on the celestial sphere. It results from the star's actual velocity through

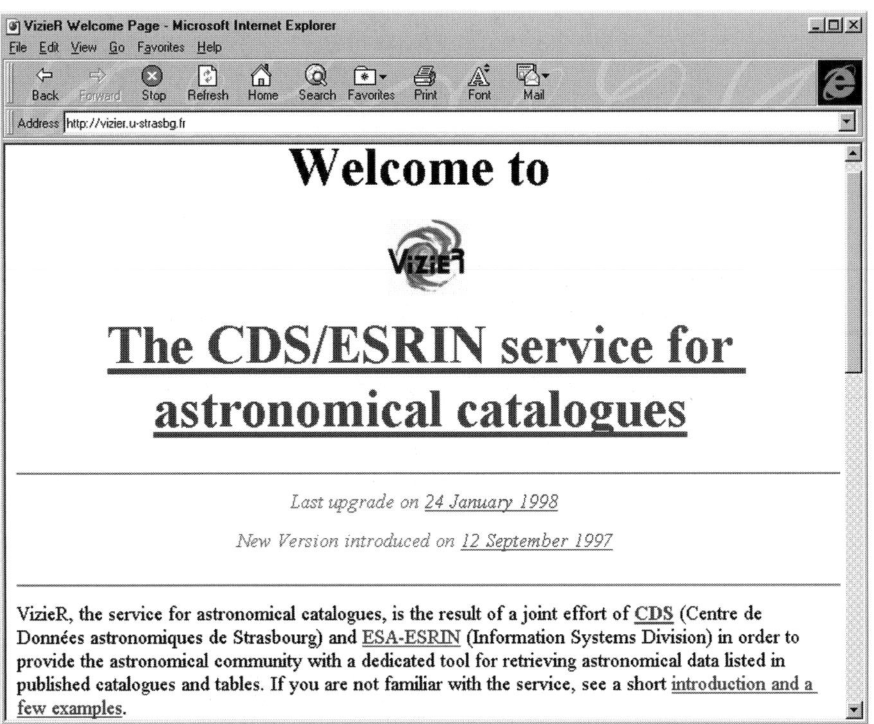

Figure 6.5. The CDS query service Homepage where VizieR is located.

space and its velocity relative to us. Everybody knows that Barnard's star has the greatest proper motion but whilst that of other stars is much smaller, taken over a long enough period of time, their proper motion becomes significant.

The Hipparcos web site provides us with a tool to display proper motions. In operation it is very similar to Sky Plot. We have to enter RA/DEC co-ordinates as before. This time we are limited to an 8 by 8 degree field of view and stars from Hipparcos only. Having displayed the field we simply choose a step size in years, 1000 years is typically a sensible figure, then click on the animate button. Keep the "tails" option on as this leaves a trail behind each star making its path more obvious. Figure 6.6 (*overleaf*) is a single screen-shot from just such an animation of the Pleiades.

What it tells us is that the stars of Pleiades do in fact share a common motion through space, strong evidence for their association (if there was any doubt!), whereas stars surrounding the cluster move in more random directions and speeds. Those surrounding stars which do match the motion of the Pleiades could well be outlying cluster members.

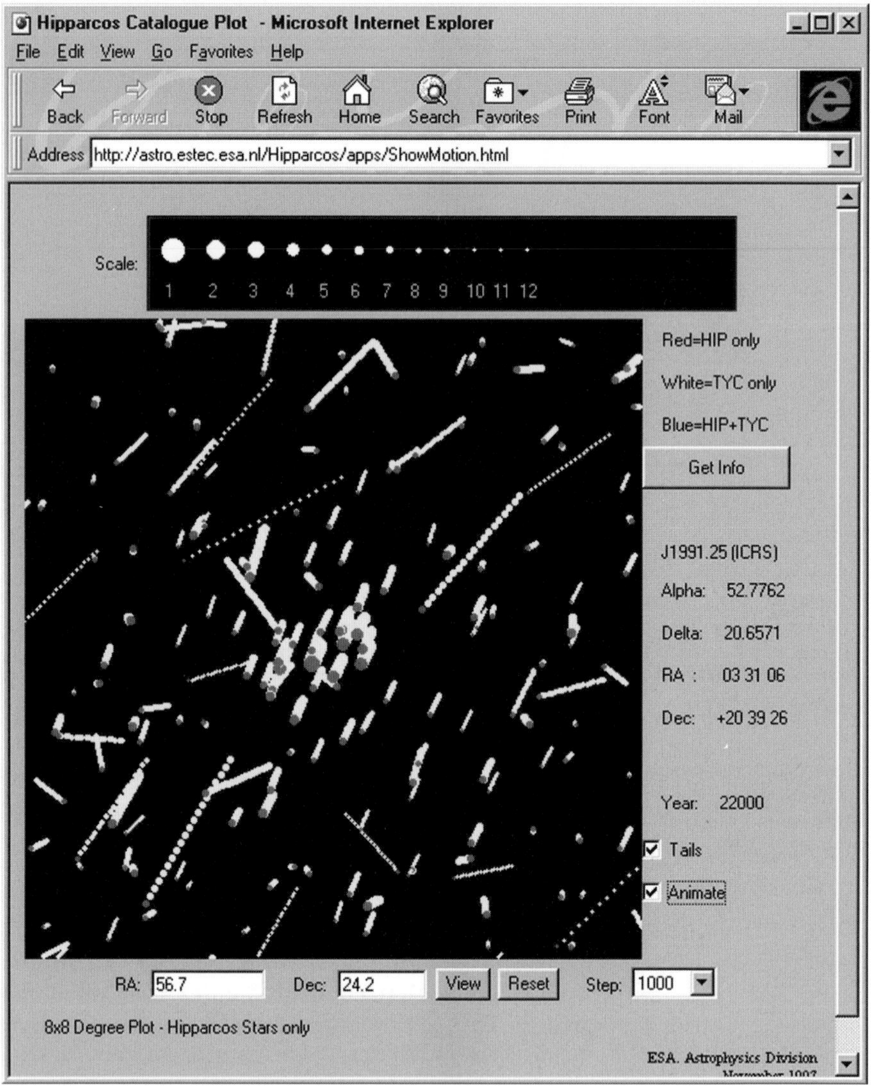

Figure 6.6. Proper motions of the Pleiades and surrounding stars.

Volume 12 of the Hipparcos and Tycho catalogues covers photometric stellar data. Hipparcos acquired an estimate of the magnitude of each star at 100–150 times throughout its nearly 4 year mission. This data is of course included when a particular star is interrogated. In addition, the web site offers an insight into fitting periods to variable stars. It is one thing to measure variability, another to determine periods accurately. A period fitting utility, known as Folded Light Curves, is available on the site with data for a few variable stars.

With this utility you can check whether the Hipparcos team has in fact found the best fit period for these stars

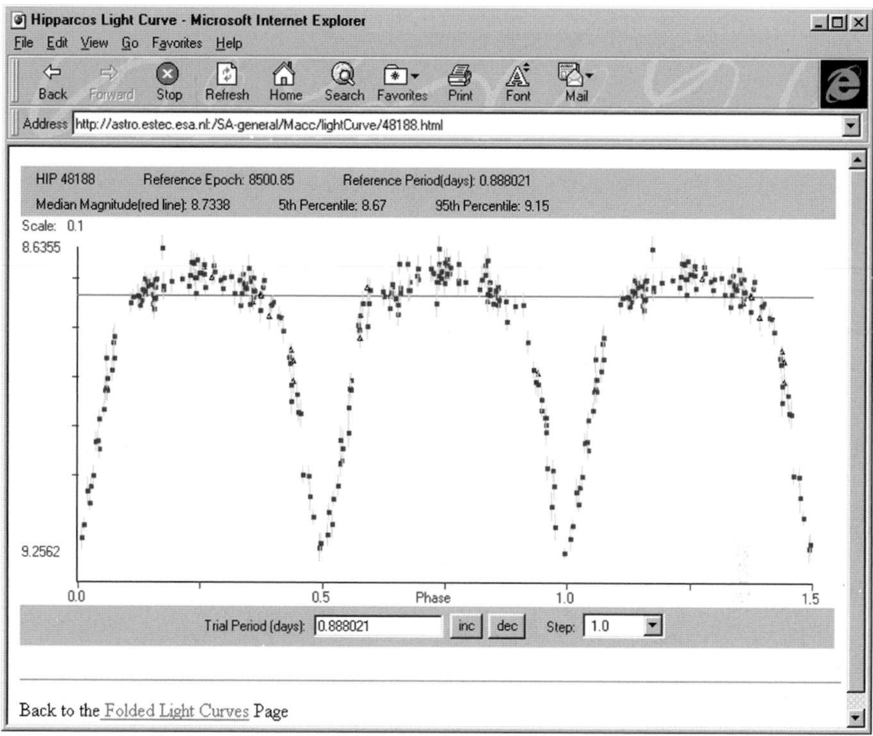

Figure 6.7. Folded Light Curves for the variable star XX Ant.

by entering your own guess and seeing the results. Figure 6.7 shows the data and light curves for the variable star XX Ant, a Beta Lyra type eclipsing binary.

Other items on the web site include 3-D stereograms of a few constellations – that for the "Plough" is included on the CD-ROM. You will need red/green 3-D glasses to see the effect. If you have only red/blue glasses then with the help of Chapter 8 you should be able to change the image to red/blue (tip – separate into RGB channels first). Also on the CD-ROM there is a 3-D animation (glasses not required) covering the motions of 48 asteroids observed during the mission. This concludes our look at the Hipparcos Homepage – what a fantastic site! It surely ranks a high place in every astronomer's favourite site list.

Herbert Raab's Astrometrica

Astrometrica is probably the leading amateur (and professional) tool for scientific astrometric and photometric

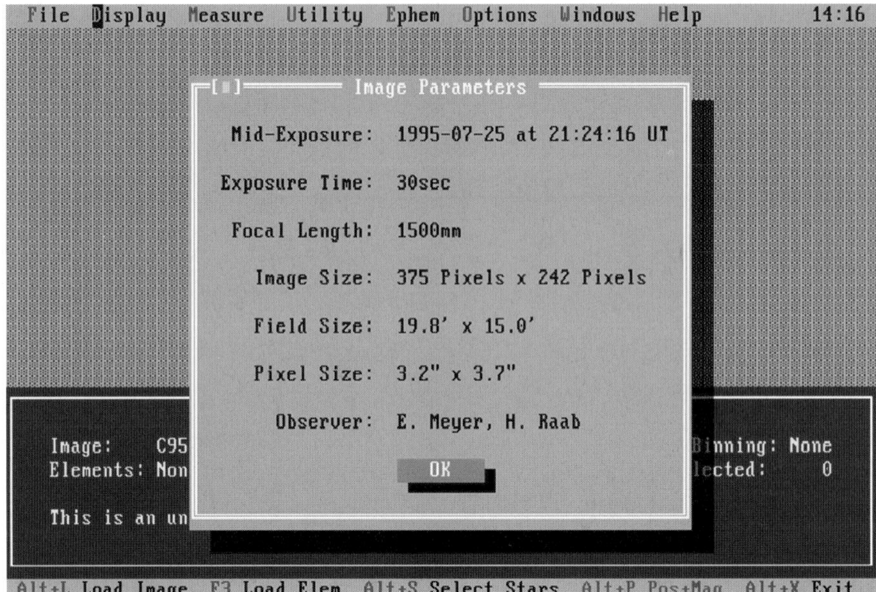

Figure 6.8. The opening screen for Astrometrica.

data reduction for CCD images. Written by Herbert Raab, an Austrian software developer, it is now in use in more than 30 different countries by many leading astronomers. It is recommended by the IAU and can produce reports directly for submitting to them. In honour of his contribution to astrometry with Astrometrica, Raab has had the minor planet 3184 named after him.

Features of Astrometrica include, a blink facility for comparing two images so that objects which have moved can be easily detected, and calculation routines for ephemerides of moving objects such as comets and minor planets. Images taken with a wide variety of common CCD cameras can be imported directly, including those from the SBIG range, the Cookbook cameras and the HiSIS range. For others it can import images in FITS (the standard astronomical file format) and TIF formats, up to a size of 2048 by 2048 pixels.

To run the program DOS is required and a VGA or VESA compatible SVGA. If you have trouble try running the program by typing:–

Astromet /NoSVGA

The program also requires a reference star catalogue. The Hubble Guide Star Catalogue is ideal but has to be obtained independently. Figure 6.9 (*opposite*) is another screen-shot from Astrometrica showing a sample image

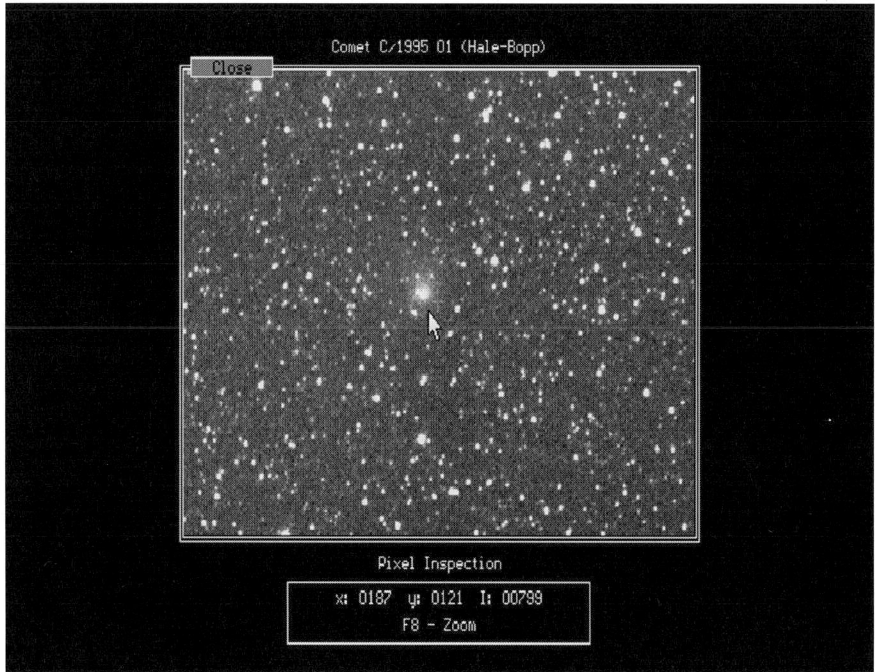

Figure 6.9. Screenshot of Astrometrica with Comet Hale-Bopp displayed ready for data reduction.

of Comet Hale-Bopp. Amateurs contributed 50 of the 63 positions for Hale-Bopp published in the IAU circulars between 24th–26th July, the vast majority of which had been calculated using Astrometrica.

An evaluation copy of the software is included on the CD-ROM but for serious users the full registered version at $25 is an absolute bargain. Included with the evaluation copy is an excellent manual which is essential reading for anyone contemplating astrometry. It includes tips on selecting reference stars and explains the mathematics behind the data reduction techniques. Once having looked at Astrometrica you will appreciate what a fitting tribute it was in naming an asteroid after Herbert Raab.

Included

on CD-ROM

COAA Orbit Determination

Suppose your existing CCD software can already calculate astrometric positions, then this little cheap and cheerful Windows (3.1 or higher) program will accept standard measurements and calculate orbital

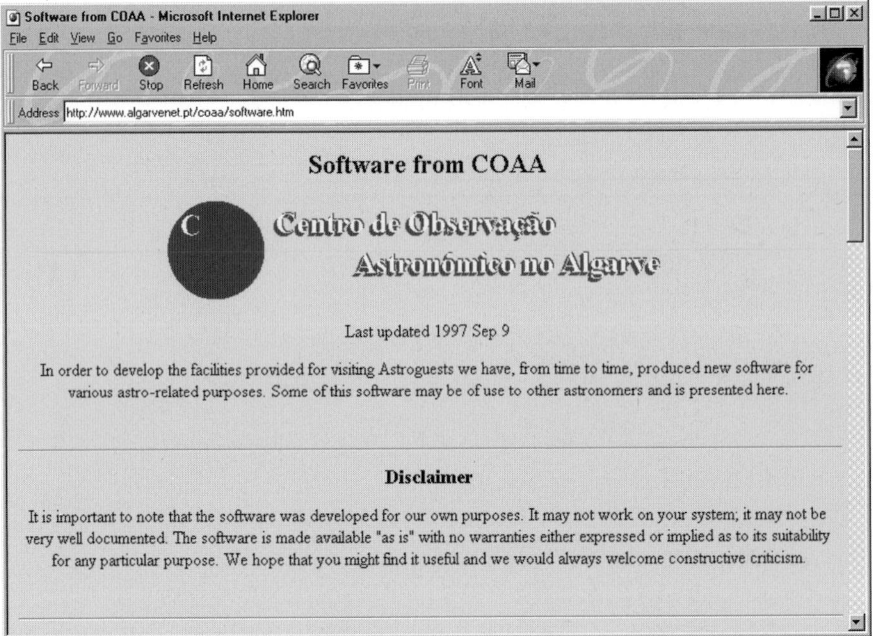

Software from COAA - Microsoft Internet Explorer
File Edit View Go Favorites Help
Back Forward Stop Refresh Home Search Favorites Print Font Mail
Address http://www.algarvenet.pt/coaa/software.htm
Software from COAA
Centro de Observação Astronómico no Algarve
Last updated 1997 Sep 9
In order to develop the facilities provided for visiting Astroguests we have, from time to time, produced new software for various astro-related purposes. Some of this software may be of use to other astronomers and is presented here.
Disclaimer
It is important to note that the software was developed for our own purposes. It may not work on your system; it may not be very well documented. The software is made available "as is" with no warranties either expressed or implied as to its suitability for any particular purpose. We hope that you might find it useful and we would always welcome constructive criticism.

Figure 6.10. The COAA Homepage – holidays and freeware!

parameters. Predictions can then be made to enable location of the object on subsequent nights. There could be nothing worse than discovering something and not knowing where to look for it on the next clear night! Written by Bev Ewen-Smith, who runs the popular astronomical holiday observatory in Portugal (known as COAA), this freeware utility, whilst not as accurate as Astrometrica, enables a quick and easy determination of which kind of object you have observed and what sort of orbit it is in. You never know, it might be going to hit earth and you would probably want to know as soon as possible!

Ideally it requires three well-spaced observations but if only two are available, it will do a polynomial extrapolation which might suffice for finding the object on the next clear night (Figure 6.11, *opposite*). It is included on the accompanying CD-ROM and is worth keeping handy, just in case you spot a mystery object!

EzPhot CCD Photometry

Included

on CD-ROM

EzPhot is a new DOS based program for automating much of the tedious reduction work in CCD observa-

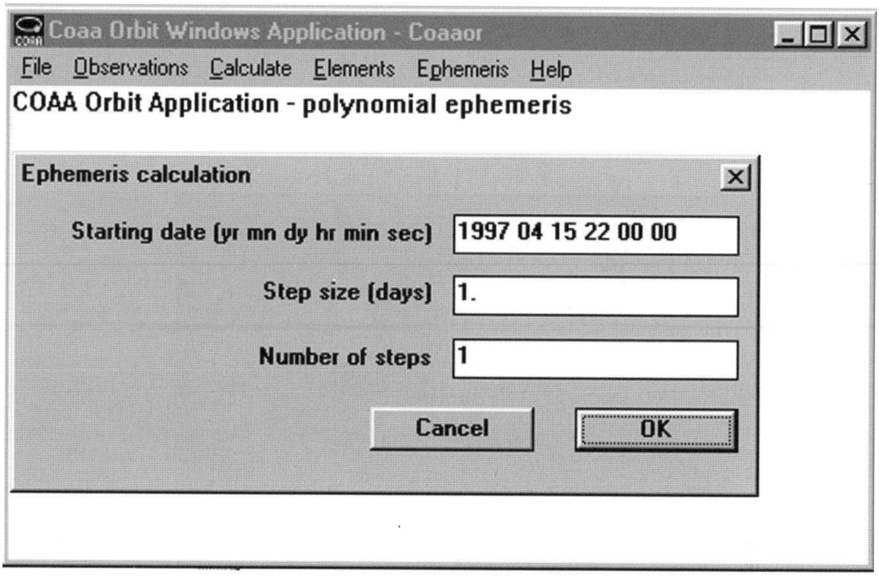

Coaa Orbit Windows Application - Coaaor
File Observations Calculate Elements Ephemeris Help
COAA Orbit Application - polynomial ephemeris
Ephemeris calculation
Starting date (yr mn dy hr min sec) 1997 04 15 22 00 00
Step size (days) 1.
Number of steps 1
Cancel OK

Figure 6.11. COAAorb program for approximate orbit calculations.

tions for stellar photometry. It will automatically co-align multiple CCD images and reduce them to magnitudes and perform differential photometry. This first version accepts FITS and SBIG file formats only.

This software was developed by Gerry Gunn who, along with Chuck Lamb, built and operates the Hanna City Robotic Observatory (Figures 6.12 and 6.13, *overleaf*). The observatory can be programmed to run unattended on its own or be remotely controlled and was featured in *Sky & Telescope*, October 1997. It is optimised for variable star photometry. For further information on this state-of-the-art control system contact Jerry Gunn:-

jgunn@mtco.com.

EzPhot.exe requires an image file as input and a flat field if available. It will find all the stars and perform aperture photometry. The files so created are then input to Register.exe which aligns them, using 6 stars. By selecting a variable, comparison and check star, differential magnitudes are output, accurate to about 0.01 magnitude. A light curve can then be plotted. The program is shareware and is limited to 30 days use. It is included on the CD-ROM.

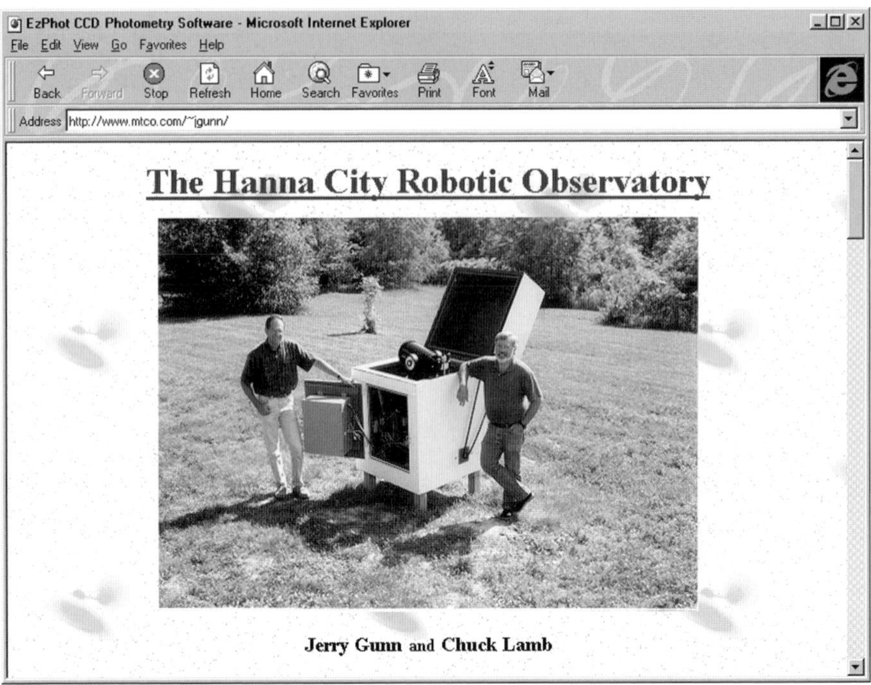

Figure 6.12. The Homepage of the Hanna City Robotic Observatory and the software EzPhot.

Figure 6.13. Screen-shot from EzPhot.

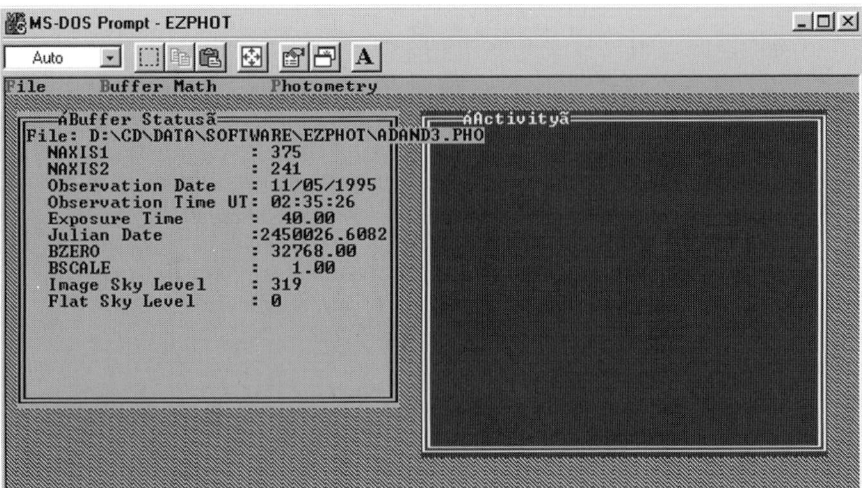

URLs Featured in this Chapter

The Hipparcos Homepage
http://astro.esctec.esa.nl/Hipparcos

The VisieR service for astronomical catalogues
http://vizier.u-strasbg.fr/

Herbert Raab's Astrometrica Homepage
http://www.planet.co.at/lag/astrometrica/astrometrica.html

The COAA Observatory software pages
http://www.algarvenet.pt/coaa/software.htm

The Hanna City Robotic Observatory Homepage
http://www.mtco.com/~jgunn

Other URLs not Referred to but Recommended

John Rogers' CCD Astrometry software Homepage
http://ourworld.compuserve.com/homepages/johnerogers

Ingrid Siegert's Astrometry Homepage
http://www.inetnow.net/~ist/astro.html

Guide to Minor Body Astrometry
http://cfa-www.harvard.edu/cfa/ps/info/Astrometry.html

Chapter 7 Images and Videos

Background

Images of astronomical wonders are what probably inspired us to become interested in astronomy in the first place and that is still true for newcomers to astronomy today. What has changed is that these awe inspiring images are much more plentiful now and, thanks to the Internet, much more accessible. I can remember the excitement of the Voyager probes reaching Jupiter and then Saturn but in those days there was no easy means for the ordinary amateur astronomer to quickly get copies of those pictures. That has all changed now. In the case of space probes, each will have its own web site with background information, mission status and, of course, the latest images, almost as soon as they have been transmitted back to Earth. We can download them and digitally process them ourselves to enhance or enlarge details, just like Mission Control used to do with those Voyager images.

A word of caution however, just because an image is published on the Internet and can be freely downloaded, the copyright still remains with the image owner. Some images are truly public domain, that is free of copyright, such as the "Public Images" released by NASA. These can be used in your own web pages for example, usually with only an acknowledgement. This explains why they appear so often in magazines and books. Other than that it is safest to assume that you are only entitled to use an image on your own PC for personal viewing. Anything else would require permission and/or a payment. NASA's "Reproduction Guidelines" covering the use of their images and emblems is included on the CD-ROM.

Astronomical Image Library

This may seem a strange place to begin this chapter. You may have been expecting spectacular NASA Hubble Space Telescope (HST) images, they come later. However, one of the problems with having millions of images available on the Internet, from thousands of web sites, is finding the particular one you want. The Astronomical Image Library is a relatively new attempt (still under construction) to provide an index to astronomical images.

It is a brilliant idea, all one has to do is fill-in, in the box provided, the object you are interested in. This can be a catalogue number, like NGC891 shown in Figure 7.1, or simply a name like "Comet Hyakutake" or the

Figure 7.1. The Astronomical Image Library Homepage and query form – see bottom left hand corner.

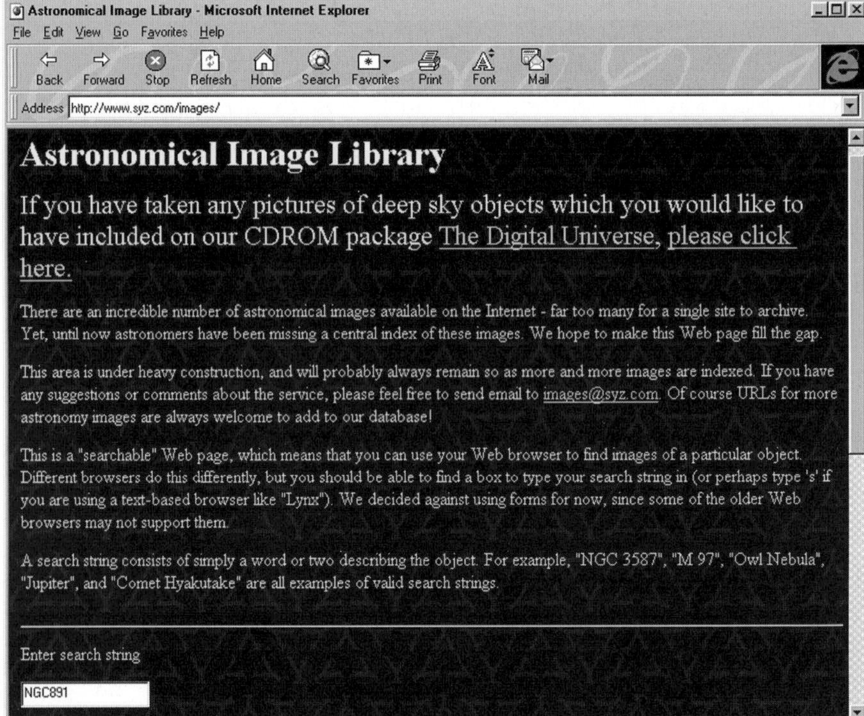

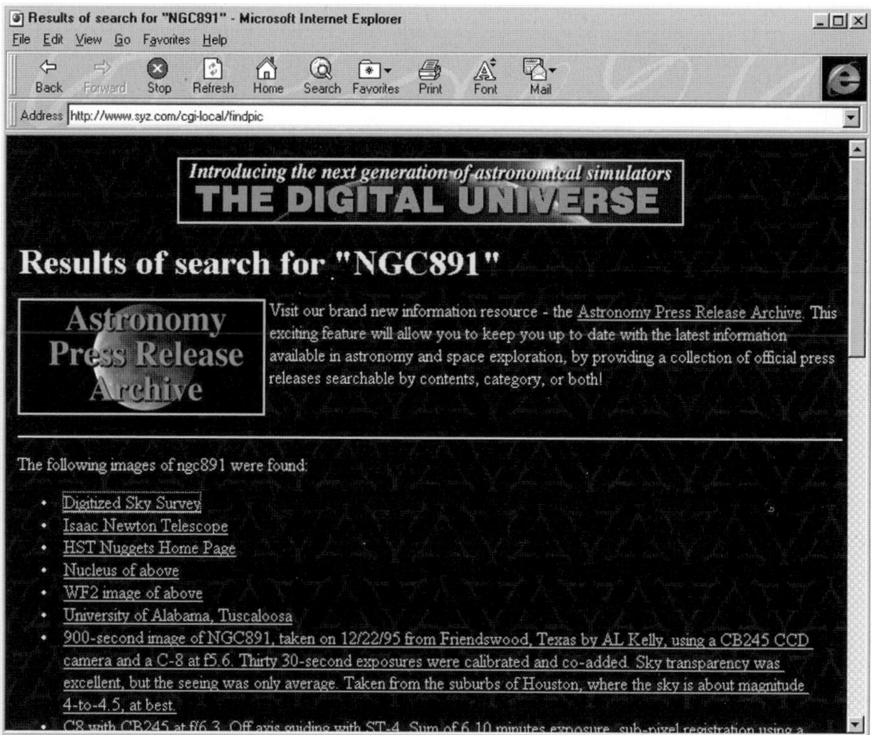

Figure 7.2. An extract of the results for the search on NGC891.

"Owl Nebula". I mentioned that the site was still under construction and as images are appearing on the Internet by the minute, it is a mammoth task to index them all and can never be complete. The service is not government funded and is provided by the authors of the software package "The Digital Universe". Having filled-in your object name or number, you will be presented with a second screen explaining this and asking for a contribution. These contributions can be offset against the purchase of The Digital Sky. Pressing "continue search" will proceed to a list of sites with images matching your query (Figure 7.2), in this example NGC891. The list includes images from the Digitised Sky Survey (see later this chapter), the Isaac Newton Telescope, the Hubble Space Telescope and even amateur ones. All that remains is to click on the hypertext link of your choice and wait patiently! Selecting the first on the list, the image shown in Figure 7.3 (*overleaf*) will appear. This can then be saved to disk if required (use the right-hand mouse button).

Overall an excellent service that will hopefully become more comprehensive over time, although funding does seem to be restricting progress. For particular topics, as

Figure 7.3. Image of galaxy NGC891 found using the Astronomical Image Library service.

we will see later, there are slicker services but none as yet offers the broad range covered by this web site.

The Digitised Sky Survey (DSS)

If it is deep sky objects you are interested in then this site could be the answer to your dreams. If you need a reference image, perhaps to check a possible supernova or new asteroid, this is the place to look. The Digitised Sky Survey is, as its name implies, digitised copies of photographs of the entire sky available on online. These original images were produced by the Palomar Observatories Schmidt Camera, down to declination −15 degrees and they are being supplemented by the images taken with the UK Schmidt Camera at the Anglo-Australian Observatory. These are deep images and the centres of bright objects are often overexposed with detail "burnt" out. They do however represent an unparalleled source of all sky images.

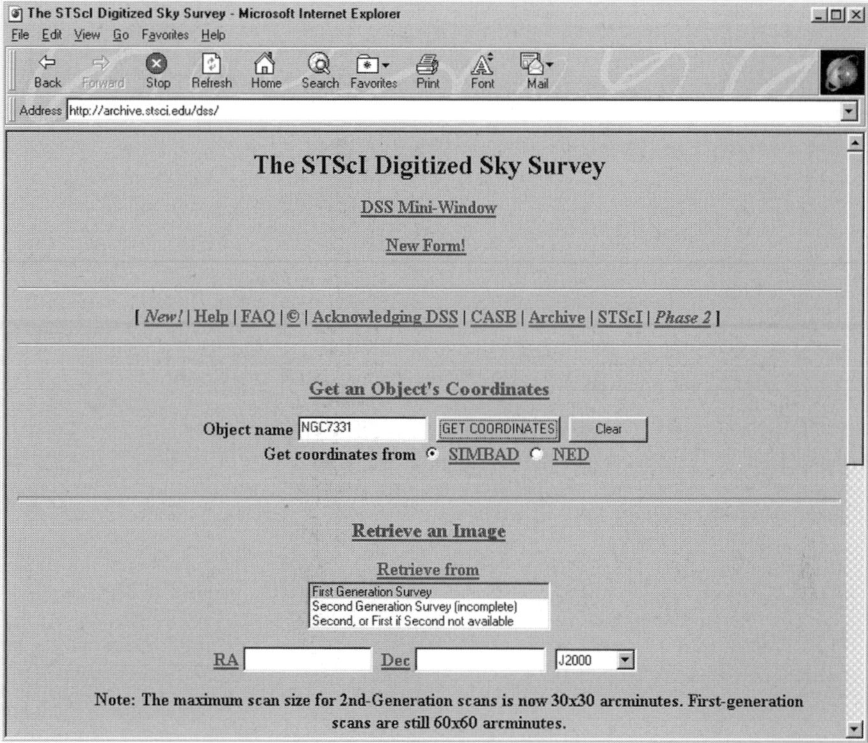

Figure 7.4. The opening screen for extracting images from the Digitised Sky Survey.

The opening screen (Figure 7.4) gets straight down to the business of obtaining an image. It is in two parts, if the co-ordinates are unknown then fill in the top half of the form and the co-ordinates of your object will be returned for you. In this example I entered the object's catalogue number, NGC7331. For finding co-ordinates two external services are offered, SIMBAD (an online database in Strasbourg) and NED (a database in Pasadena). SIMBAD contains a comprehensive list of all kinds of objects whilst NED concentrates on objects outside our galaxy. Either way, the co-ordinates should be returned fairly quickly, if not it could mean that the link is down and you will have to try later.

Assuming that the co-ordinates are returned, or we knew them already, then we can proceed. We can select images from the first generation survey or the still ongoing second generation survey, or a combination of both. Image scales are not the same however, the second generation ones being scanned at a different resolution. Other options for us to fill-in include the image size required, up to a maximum of 60 arcminutes square, and the image format we require, GIF or FITS. GIF is probably the better option unless you have

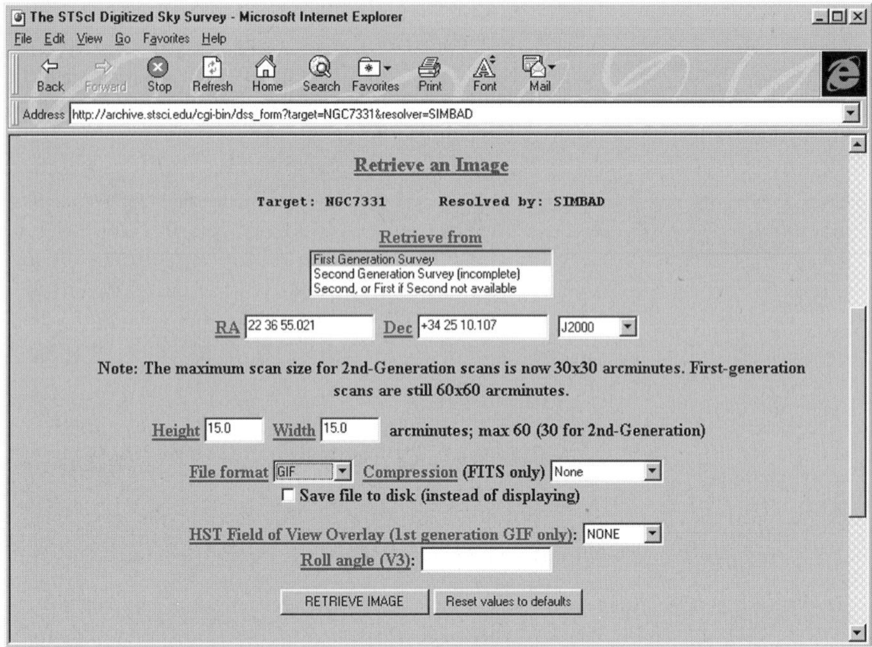

Figure 7.5. Completed form for obtaining an image.

Figure 7.6. Image returned of galaxy 7331 as requested. Note it is slightly off centre. Co-ordinate error?

software that will display the FITS alternative (see Chapter 8). There is also an option for an HST field of view overlay. Depending on your choice of size, in a few minutes an image of the area selected will be displayed (Figure 7.6).

Included on CD-ROM

The question often asked of the DSS is, is it possible to submit a batch of queries? The answer unfortunately is no. Batch requests would simply swamp the service, so I am afraid it is only one at a time. However several of the popular Messier objects have been downloaded and are included for reference on the accompanying CD-ROM.

NASA's Public Images Photo Gallery

The sections of this chapter that follow can all be located through this NASA web site and it is definitely the place to start. It is an attempt to bring together as many of NASA's images as possible and is extremely comprehensive. As well as the Hubble Space Telescope,

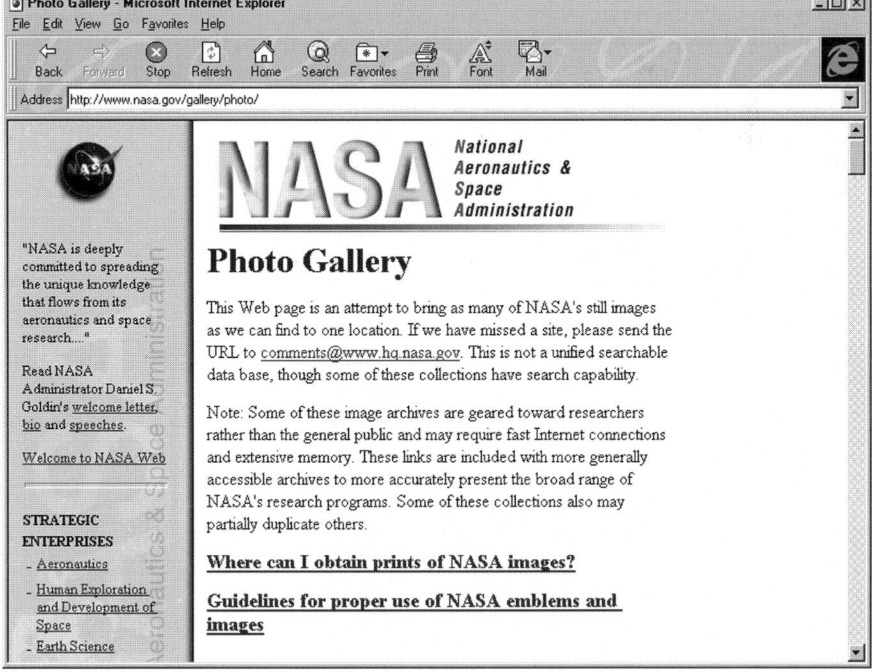

Figure 7.7. The gateway to NASA's Photo Gallery of Public Images.

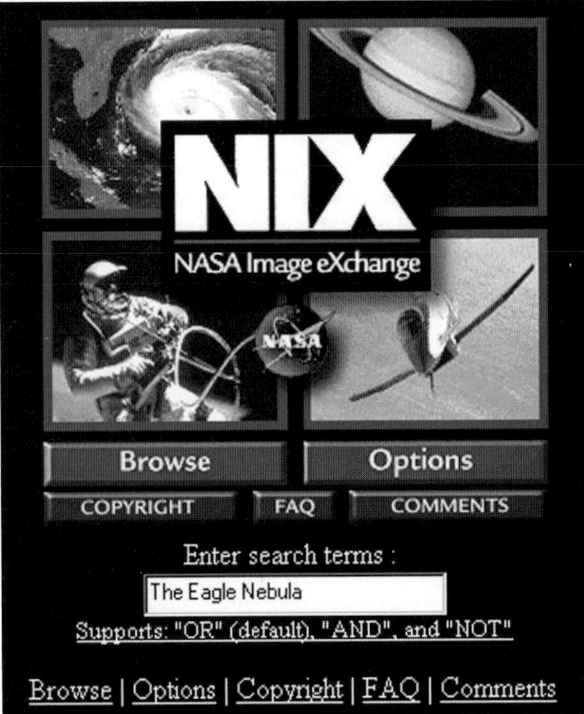

Figure 7.8. The NIX image search facility on the NASA Photo Gallery Homepage.

the index covers Astronomy, the Earth (including oceanography and ozone studies), the Solar System and the Space Shuttle. There is a search facility called the NASA Image Exchange (NIX) and it will rapidly find and locate the image of your choice.

We will now look at some of the sites located either by NIX or direct from the links on NASA's Photo Gallery homepage. Remember to check the "Guidelines" covering the use of these images and you will find them on the accompanying CD-ROM.

The Hubble Space Telescope (HST)

If we use the NIX search facility for a deep sky object, such as the Eagle Nebula, inevitably we will be directed to these homepages. The Hubble Space Telescope (HST) web site is home to some of the most spectacular images ever taken. They are the ultimate in high resolution

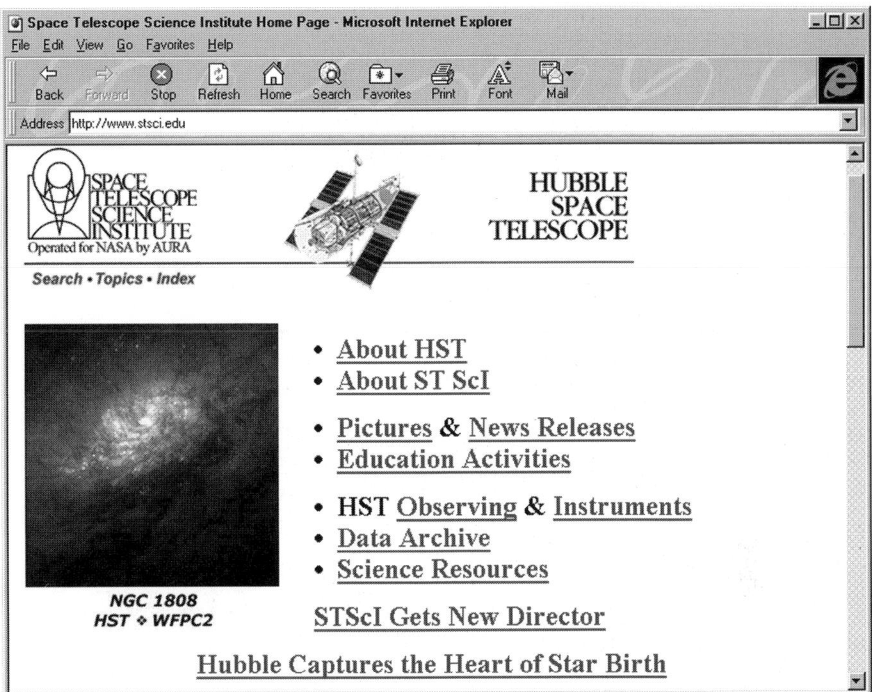

Figure 7.9. The Hubble Space Telescope Homepage – the least spectacular part of the site!.

imagery and whilst these pictures were undoubtedly taken for their scientific value, they provide a high public profile for NASA. The only real problem is the sheer volume and size of the images produced. Originally they were issued in both high resolution JPEG format and lower resolution GIF format. The latter had the benefit of supporting embedded text, and usually included with the image was a descriptive text. So, providing you had a GIF viewer capable of showing the embedded text and you realised it was there, you could read the background information. However, this was a far from convenient way of learning about the image so now NASA also issues its public images in Acrobat portable document format (PDF). This bundles graphics and text together in a newsletter style format which can be viewed on any computer with an Acrobat reader e.g. PC, Macintosh, or even Unix. This format had long been used by the US Tax authorities for their tax forms – employing it for astronomical images is a more welcome use!

Public images and press releases are organised by year (Figure 7.10, *overleaf*), plus a few other topics such as the modestly named HST's Greatest Hits and

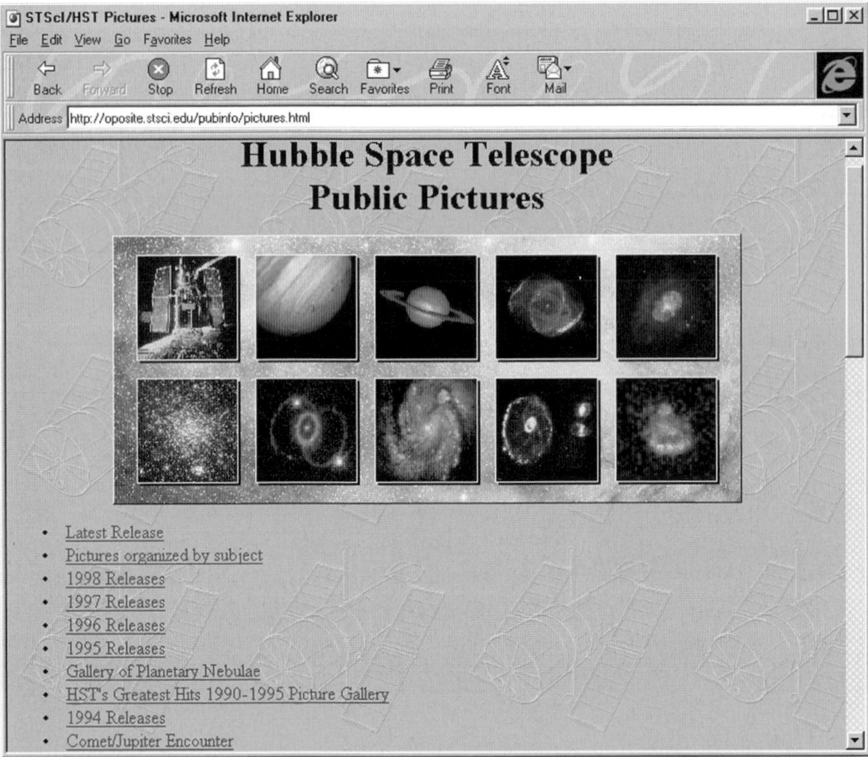

Figure 7.10. Public Pictures at the HST web site.

the spectacular Gallery of Planetary Nebulae. Some of these links contain thumbnails (miniature images), making the selection of the correct one almost foolproof. Having selected an image and Figure 7.11 (*opposite*) is a typical one for NGC1808, we are presented with a list of options. We can display a text caption, the image in two resolutions, the highest at 300dpi, or an Acrobat PDF document with both text and graphics. File sizes are quoted to give us an idea of the time they will take to download. Note, images are now usually available in JPEG format which, although it incurs an image degradation, does support full colour.

NASA also publishes on this site several videos and animations. These tend to be big files, several Mbs usually, and they are in MPEG format with a few in Quicktime format. These deserve to be better known, I suspect their sheer size has prevented their widespread dissemination. However, many of the images and videos are included on the CD-ROM so enjoy them at your leisure, free from the worries of high download costs!

Included

on CD-ROM

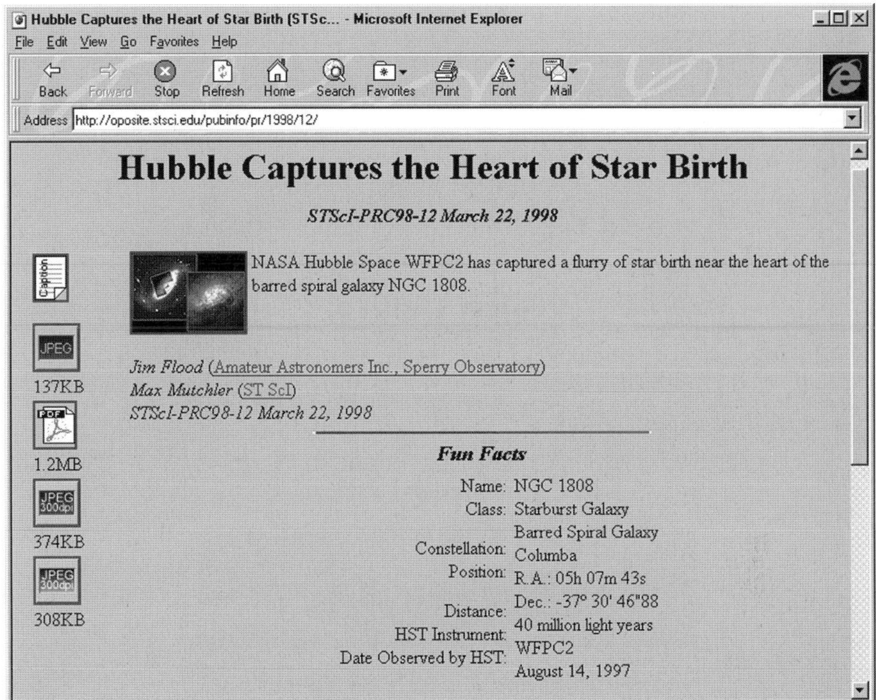

Figure 7.11. Results of selecting NGC1808 – note the options for images and captions with file sizes quoted.

NASA/JPL's Project Galileo

As referred to in the introduction to this chapter, space probes invariably have their own Homepages. Launched in 1989, NASA's Galileo spacecraft arrived at Jupiter (after a very circuitous 8 year journey) in December 1995. This route enabled it to image Earth and the asteroids Gaspra and Ida en route. Because of data transmission problems, Galileo can only send its data back to Earth very slowly. As a result it was some time after its encounter with Ida that Galileo was discovered to have imaged an unknown satellite of it. This 1 kilometre wide body has since been named Dactyl.

Once in orbit around Jupiter it returned superb images of the planet and its moons. Its mission was due to end in 1997 but has been extended for a further two years to concentrate on Europa with two flybys of Io. Several of Galileo's images taken on its journey to and from its orbit around Jupiter have been included on the CD-ROM.

Included

on CD-ROM

Figure 7.12. The Project Galileo Homepage.

The NASA Mars Missions: Pathfinder, Global Surveyor & Surveyors '98 and 2001

On 4th July 1997, NASA's Mars Pathfinder landed safely on the surface of Mars, the first spacecraft to do so in over 20 years since the Viking probes. This, the first of several NASA missions to return to the red planet, was a huge success. It landed in Ares Vallis, a rock strewn plain which showed some evidence of ancient flooding. The day after it touched down, a miniature 6-wheeled rover named Sojouner was deployed and spent the next month visiting and imaging surrounding rocks. Boulders with names such as "Yogi" and "Barnacle Bill" became almost as well known as their cartoon forebears, whilst on the horizon, features such as

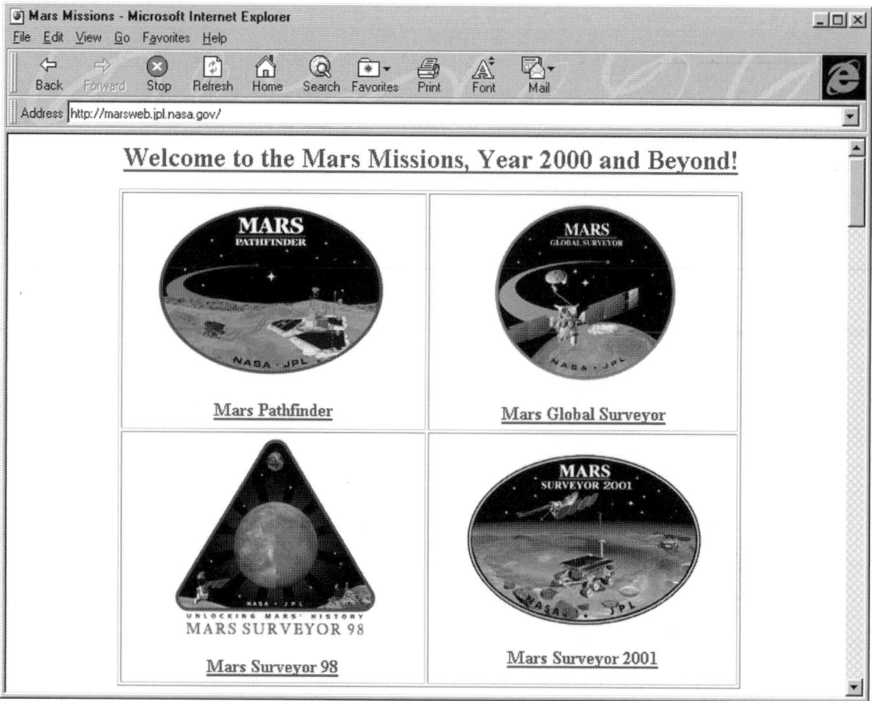

Figure 7.13. The NASA Mars Missions Homepage.

"Twin Peaks" became equally famous. As a mark of respect to the late Carl Sagan, the Pathfinder lander was renamed the Carl Sagan Memorial Station by NASA.

Following Pathfinder there was the Mars Global Surveyor which entered orbit around the planet in September 1997. After the slow and painstaking process of aero-braking into a circular orbit, the probe will begin its task of imaging the Martian surface. Some early test images have been returned including one to end, once and for all, the controversy over the "Martian Face" imaged by Viking.

Included on CD-ROM

A feature of the Pathfinder/Sojourner images has been the number of 3-D ones created and both conventional and 3-D images are included on the CD-ROM. Should any new Surveyor images be returned, they will also be included, so check the CD-ROM for the latest position. Missions to follow these two have already been programmed for 1998 and 2001. Next time the lander will touch down at one of the poles and look for water. We have some exciting times ahead.

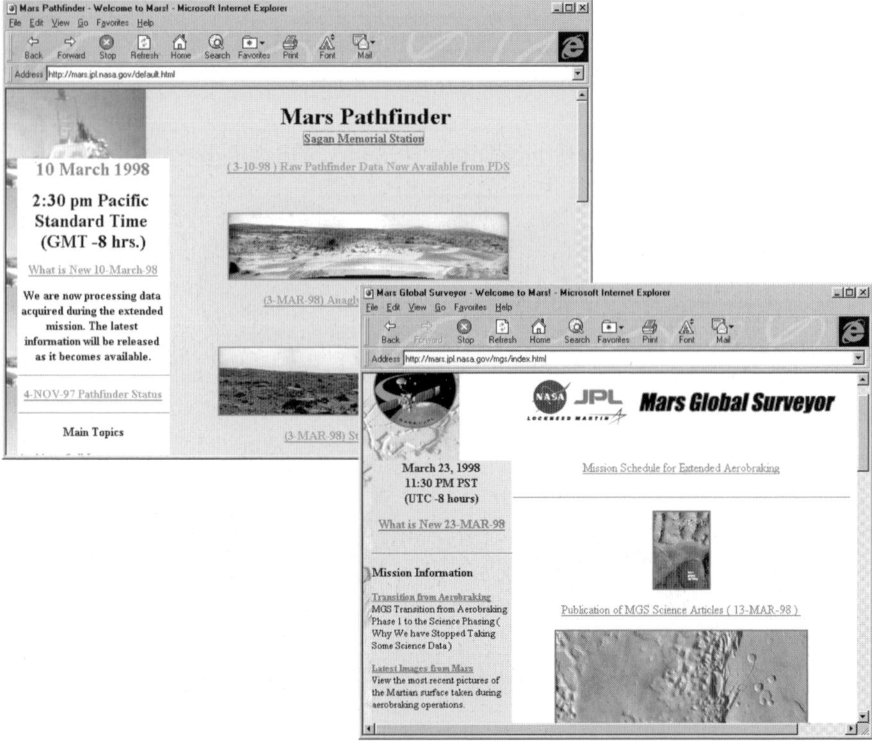

Figure 7.14. Images from the Mars Pathfinder and Global Surveyor.

NASA's Earth from Space

Finally after our tour of the Universe and Solar System, we must not forget that cameras can look down as well as up. NASA's Earth from Space web site does just that! Selected photographs from NASA's database of over 250,000 images of Earth are being put online. These images have been taken by Astronauts and unlike Earth Resources satellite images are spontaneous views, often at oblique angles, giving us a real indication of what the Earth looks like from orbit.

Provided on the site is a search engine and images are sorted by categories which include cities, landscapes, regions, water habitats, weather etc. Unfortunately for the UK, the Shuttle's orbit does not often come

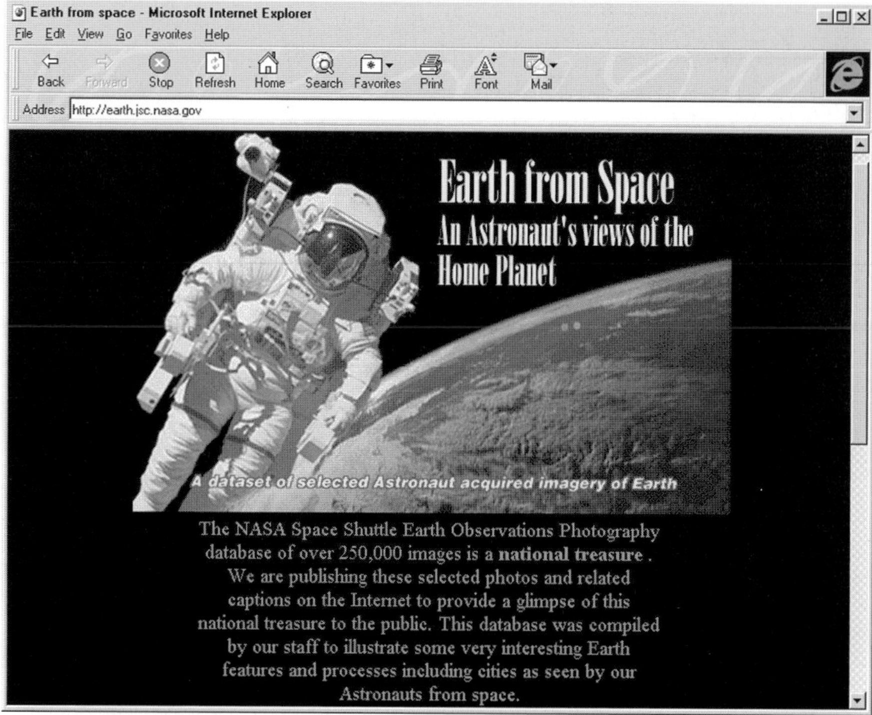

Figure 7.15. NASA's Earth from Space Homepage.

sufficiently far north to image us, so there are very few of the British Isles. As an alternative to searching by category, a World Map can be viewed with image locations marked on it (Figure 7.16, *overleaf*). It is a simple operation to click on the area of interest and get your image. Images are available in two resolutions, the highest resolution ones being 5000×5000 pixels which even in compressed form are around 5 Mb in size! Low resolution ones at typically 640×480 pixels are more practical at around 100k each.

Included on CD-ROM

A small selection of low resolution images is included on the CD-ROM.

As more images are added to this site its usefulness will increase. It is therefore well worth keeping on your list of sites to visit.

URLs Featured in this Chapter

Astronomical Image Library
http://www.syz.com/images

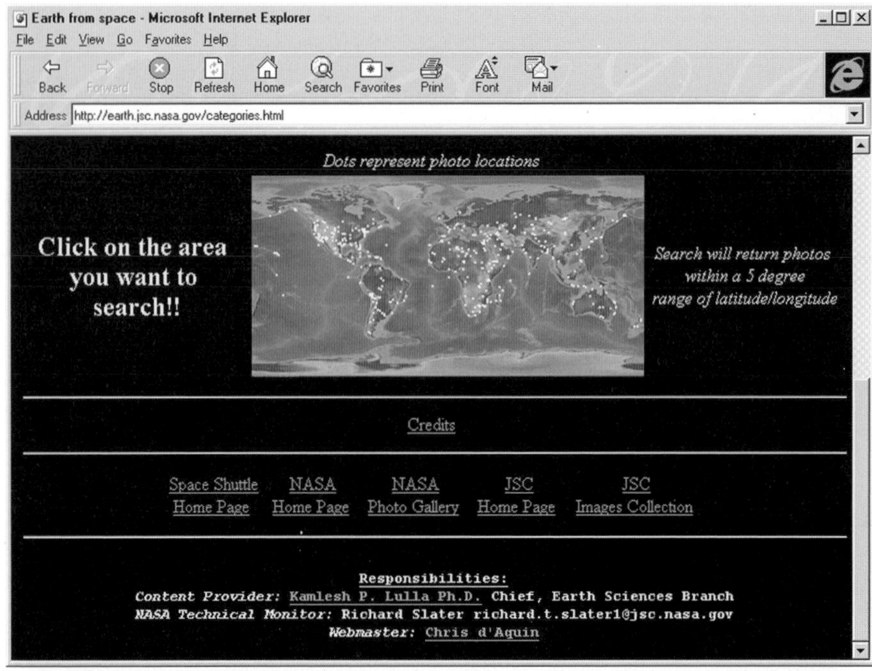

Figure 7.16. Searching for images by viewing a World map.

Digitised Sky Survey (DSS)
http://www.stdatu.stsci.edu/dss

NASA's Photo Gallery & NIX Search Facility
http://www.nasa.gov/gallery/photo/

Hubble Space Telescope's Public Pictures
http://www.stsci.edu

Project Galileo
http://www.jpl.nasa.gov/galileo

NASA's Mars Missions
http://marsweb.jpl.nasa.gov

NASA's Earth from Space
http://earth.jsc.nasa.gov

Other URLs not Referred to but Recommended

Clementine Space Mission to Saturn
http://www.nrl.navy.mil/clementine

Anglo-Australian Telescope Images
http://www.aao.gov.au/images.html

WIYN Telescope Images
http://www.noao.edu/wiyn/wiynimages.html

European Southern Observatory Images
http://www.eso.org/epr/press-releases-list.html

Big Bear Solar Observatory
http://bigbear.caltech.edu

The Virtual Observatory,
not original images but a good route to finding them
http://www.stud.ifi.uio.no/~mikkels/astropics.html

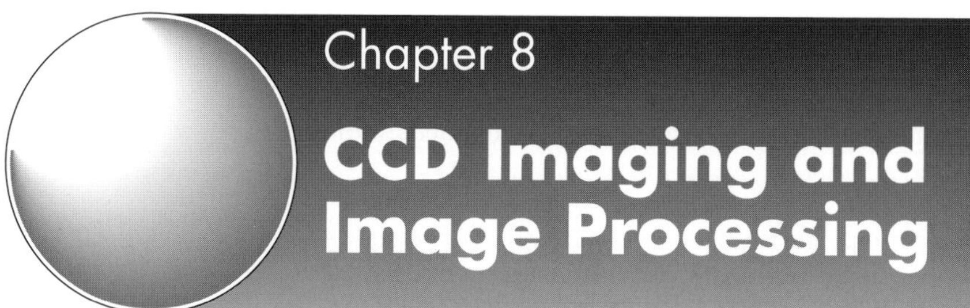

Chapter 8
CCD Imaging and Image Processing

Background

The advent of affordable CCD (Charge Coupled Device) cameras has transformed amateur astronomy. Not only are they around ten times more sensitive to light than conventional film, they deliver a digitised image which is easy to enhance and analyse using a personal computer. Gone are the days when dark room wizardry was a prerequisite for successful imaging. We have already seen in Chapter 6 that real scientific work, such as astrometry and photometry, is now routinely carried out using these devices. Even for the amateur, with more modest ambitions, CCD cameras now enable spectacular images to be captured from urban back gardens, where light polluted skies have rendered conventional observing almost impossible.

There are many sites on the Internet, where amateurs display their fantastic images, but perhaps too many of them concentrate on showing images rather than explaining the "hows and wherefores". For further details on using these cameras I would not unnaturally recommend to you the book, also in the Practical Astronomy Series, *The Art & Science of CCD Astronomy* edited by myself, where experts from around the world pass on their tips and methodology for successful imaging. Many of the contributors have their own web pages and their URLs are quoted at the end of this chapter.

In this chapter we concentrate on software with two aims in mind. Firstly for those wishing to try image processing before making the investment in a camera and secondly for those who have a camera and are interested in image processing. For the former, I would suggest experimenting with the images included on the CD-ROM, such as the Digitised Deep Survey ones. There is much scope for improving these as little or no enhancement has taken place on them. However we first look at the home of the Cookbook Camera, where Richard Berry has brought together all manner of tips and improvements for these home-made CCD cameras.

The Cookbook CCD Camera Homepage

The Cookbook CCD camera is in fact two cameras, described in the book *The CCD Camera Cookbook* by Viekko Kanto, John Munger and Richard Berry, published by Willmann-Bell. The cameras differ only in the size of chip they use. They can be built for around $500 (£300) and feature Texas Instruments CCD chips rather than the camcorder derived chips more usual in cheaper cameras. There can surely be nothing more

Figure 8.1. Richard Berry's Cookbook CCD Camera Homepage.

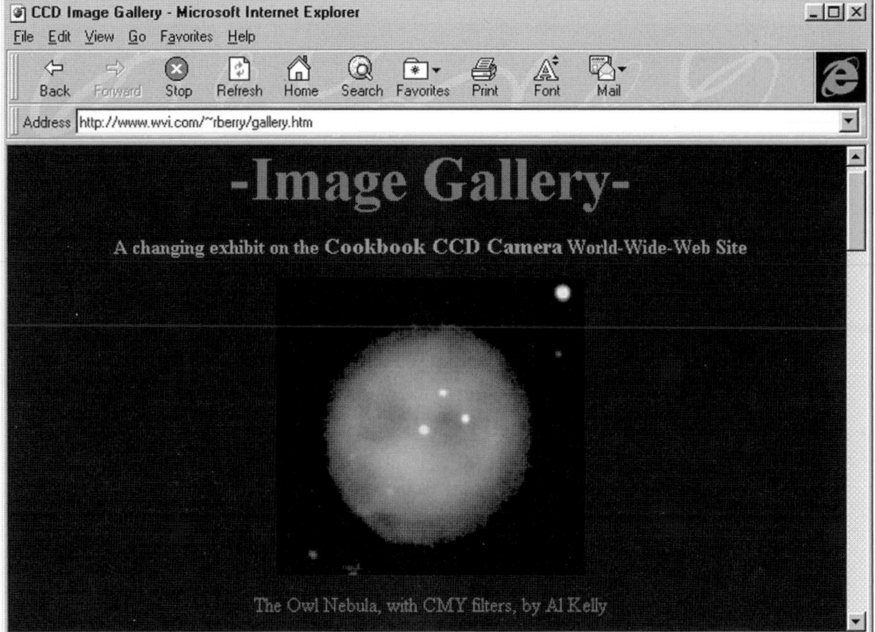

Figure 8.2. The Cookbook camera Image Gallery with many stunning images.

satisfying than imaging the wonders of the universe with a camera you made yourself.

The CCD chips used in the cameras are either the TC211 with 192×165 pixels or the larger TC245 with 378×242 pixels. A feature of the latter is its large pixel size of 17×20 microns. Large pixels equate to faster imaging and are ideal for the bigger telescope, say with focal lengths of around 2 metres. Many stunning images have been taken with these cameras as a perusal of the image gallery (Figure 8.2) will reveal. A particular advantage of the camera is that only a cheap old 386 PC is required to operate it. An important factor if you are imaging in less than ideal situations, such as in the open air or in an uninsulated observatory.

Included on CD-ROM

If this has whetted your appetite, included on the CD-ROM is the Frequently Asked Questions (FAQ) page from this web site.

Included on CD-ROM

FITSview Astronomical Image Viewer

This little utility for Microsoft Windows (3.1 or higher) enables astronomical image files in the FITS format

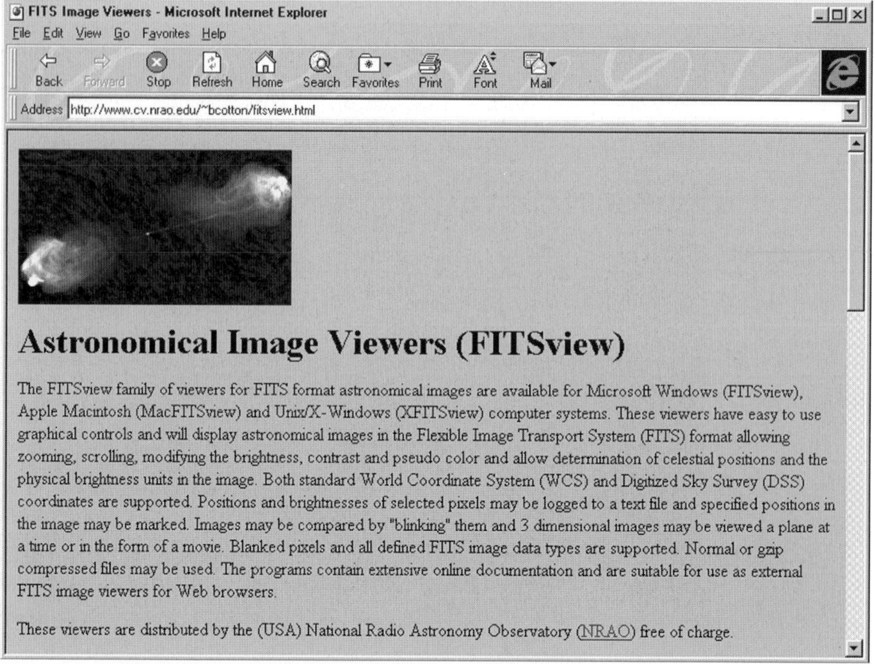

Figure 8.3. The FITSview Homepage.

to be read and displayed. It is distributed free of charge by the National Radio Astronomy Observatory (NRAO). The FITS format is the astronomical standard for files (it is not just for images) and is used by professional astronomers world-wide plus increasingly by amateurs too. You may recall that one of the options on the Digitised Sky Survey web site was to select the resultant image to be in FITS format. It has the advantage over the more normal PC formats such as TIF, GIF and BMP of preserving the full dynamic range in an image. Even amateur CCD images can have 16,000 shades of grey in them – detail that would be lost in simpler formats. FITS also allows the transfer of images from one proprietary package to another. FITS images can now be compressed by a process known as gzip and FITSview can handle these too. If you have an older image processing package then these files could be unreadable without using FITSview to uncompress them first.

Although described as a viewer, some image processing is possible. Functions are provided for altering the brightness and contrast (using slider bars) and false colour palettes are available. The latter, whilst producing gaudy results, can be useful for revealing faint detail. Zooming and panning and co-ordinate determ-

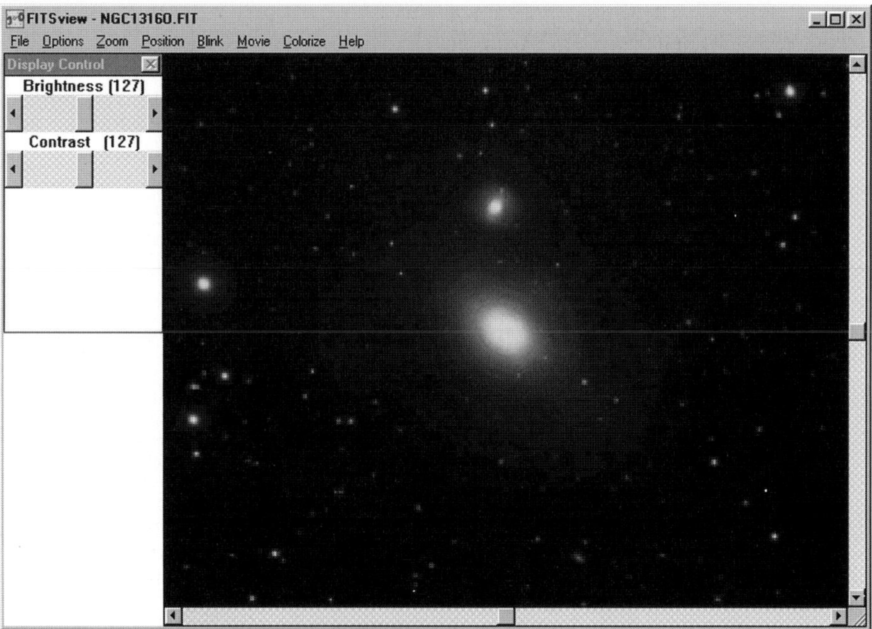

Figure 8.4. FITSview in action, zoomed in to see the galaxy NGC1316.

ination are also available. Another feature is blinking, whereby two images can be rapidly displayed so that any object which has changed or moved becomes readily apparent. Files can be saved in BMP format (with brightness and contrast changes made) for inputting into other Windows programs. The software together with an excellent help file is included on the CD-ROM. Use the Setup.exe utility to install it but read the Readme.txt file first. Two demonstration images NGC1316R.FIT and NGC1316O.FIT are included – try blinking them! Other versions of FITSview for Unix and Macintosh computers are available from the web site. An excellent little package you will want to keep in your armoury.

Included

on CD-ROM

Christian Buil's WinMiPs

Anybody who has read anything on CCD imaging will have heard of Christian Buil. He, probably more than any other person, was responsible for the CCD revolution for amateur astronomers and, whether they realise it or not, we all owe a debt of gratitude to him. His

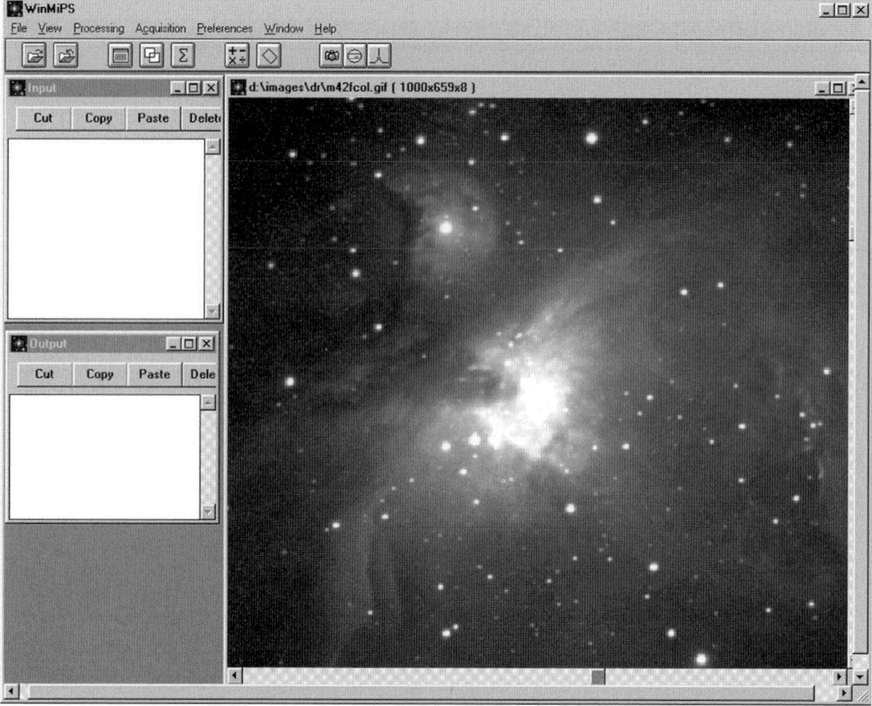

Figure 8.5. WinMiPs with an image of the Orion Nebula.

book, *CCD Astronomy*, first published in French in 1989, later translated into English and published by Willmann-Bell, inspired most early CCD camera makers and imagers. A programmer by profession he, along with colleagues, developed the well known MiPs image processing software. That DOS based package has virtually every function you could possibly require from simple image calibration to sophisticated maximum entropy deconvolution routines. It has one major drawback however. Its user interface is very user-unfriendly and only the most determined will be able to master it.

No doubt in answer to the need for a more user-friendly image processing package, Christian subsequently developed the Windows based WinMiPs. This has many of the features of MiPs, although some of the exotic deconvolution routines are absent, but it has a more friendly "point and click" format.

In MiPs, file names and their paths have to be laboriously typed-in (assuming you can remember them correctly), whereas in WinMiPs you can simply highlight files in displayed lists. There are too many features in WinMiPs to list them all. It contains all the necessary tools to calibrate images and increase their

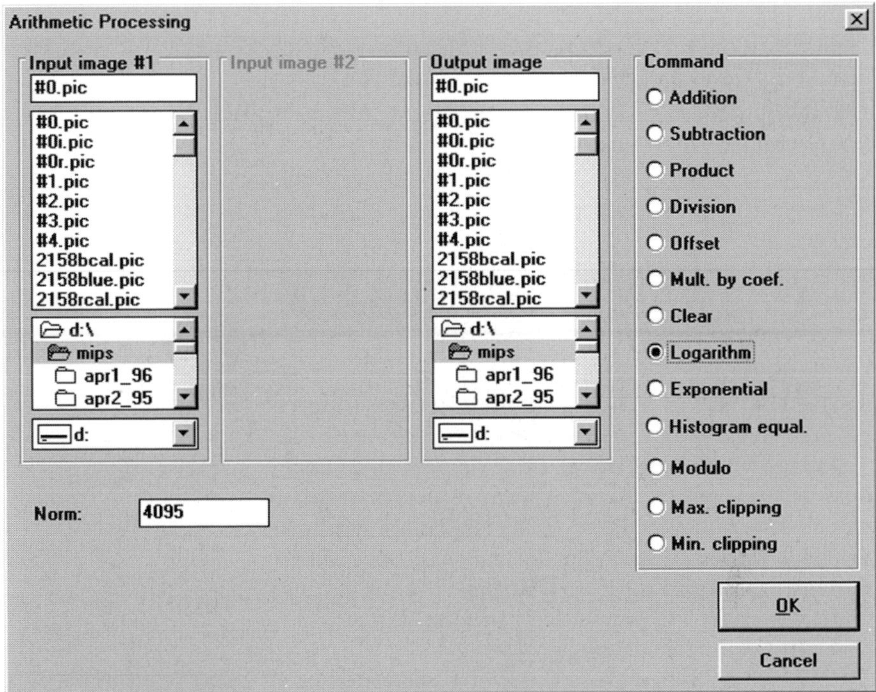

Figure 8.6. File handling and processing with WinMiPs.

Figure 8.7. Image import options for WinMiPs.

quality with commands for logarithmic scaling (useful for revealing faint detail) and the essential unsharp masking, the digital equivalent to that which the master astrophotographer David Malin uses. "Mosaic" is a useful utility for assembling a large image from a series of overlapping smaller ones. Essential for such objects as M31 which are usually too big to image in one go. It also has functions for image analysis for both astrometry and photometry.

It uses its own internal image format (PIC) but has extensive import facilities for common formats such as FITS, TIFF, BMP, GIF, PCX, TARGA, JPEG and even free format (see Figure 8.7).

Import
Free FITS TIFF BMP GIF PCX TARGA JPEG
Image name: d:\images\dr\m42fcol.gif
24 bits -> 8 bits
RGB conversion
Dimension X: 1
Dimension Y: 1
Header length (bytes): 0
OK Cancel

Version 1.5 of WinMiPs is now freeware and is included on the CD-ROM together with extensive help files and sample images. To run, copy the folder to your hard drive and click-on Winmips.exe. You will need to set the correct path to the image files included (see preferences option). An excellent way to see if image processing is for you!

Quantum Image

Included on CD-ROM

Let us suppose you have already taken some CCD images but they leave, how shall I put it, something to be desired. It is not just the original Hubble Space Telescope which experienced imaging problems! Poor guiding, atmospheric turbulence and optical inadequacies all affect image quality. Quantum Image is a new Swedish image processing program which specialises in image restoration. The shareware version runs under Windows 95/98 and is very slick but is restricted to image sizes of 256×256.

Currently included are image restoration methods using the popular Lucy-Richardson (referred to paradoxically as Richardson-Lucy) algorithm and fast fourier transforms (FFT) – see Figure 8.9 (*opposite*).

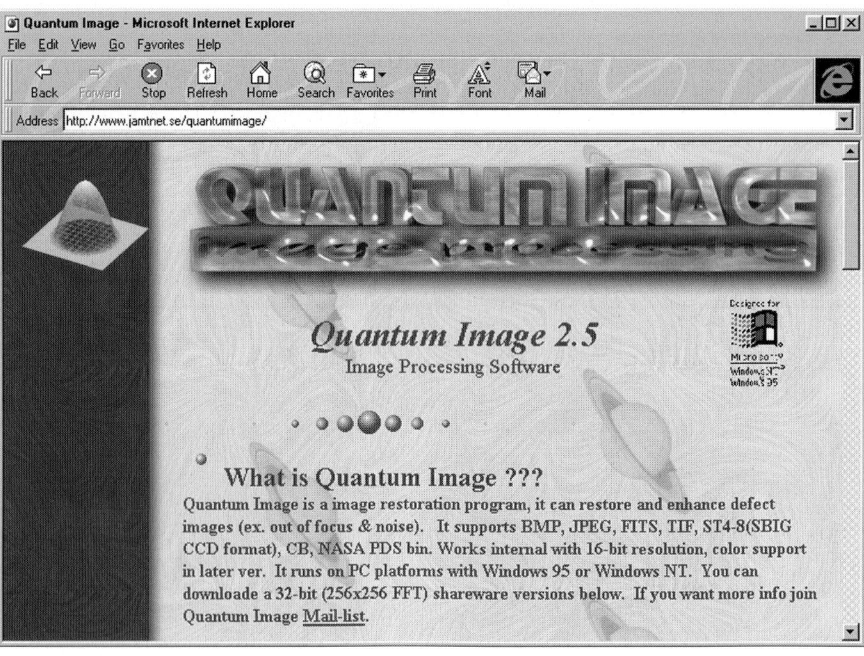

Figure 8.8. The Quantum Image Homepage.

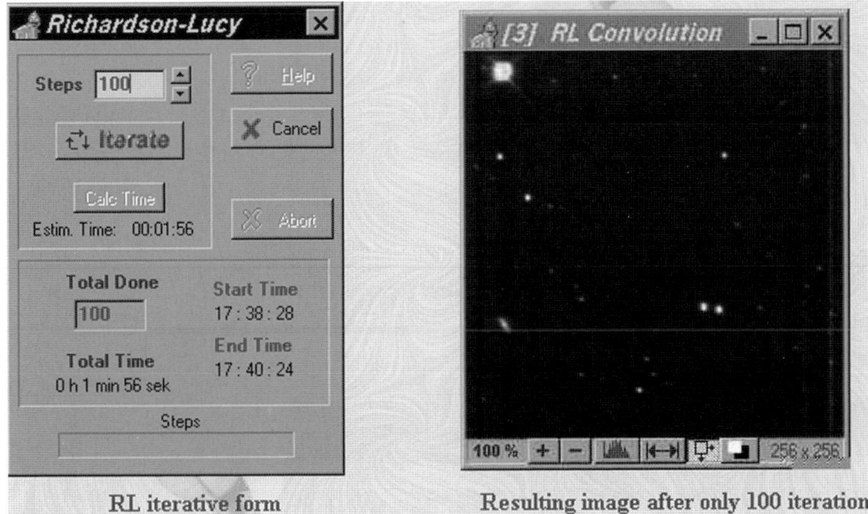

Figure 8.9. Image restoration using the Lucy-Richardson (Richardson-Lucy) method.

These work by extracting data from a typical star in the image (Figure 8.10). Then, by an iterative process, restoring it and the rest of the image to what it probably should look like. That is, as if everything had been perfect during the image taking process. They tend to work best on images with a good signal to noise ratio.

Figure 8.10. Extracting data (PSF) for a typical star in the image so it can be restored.

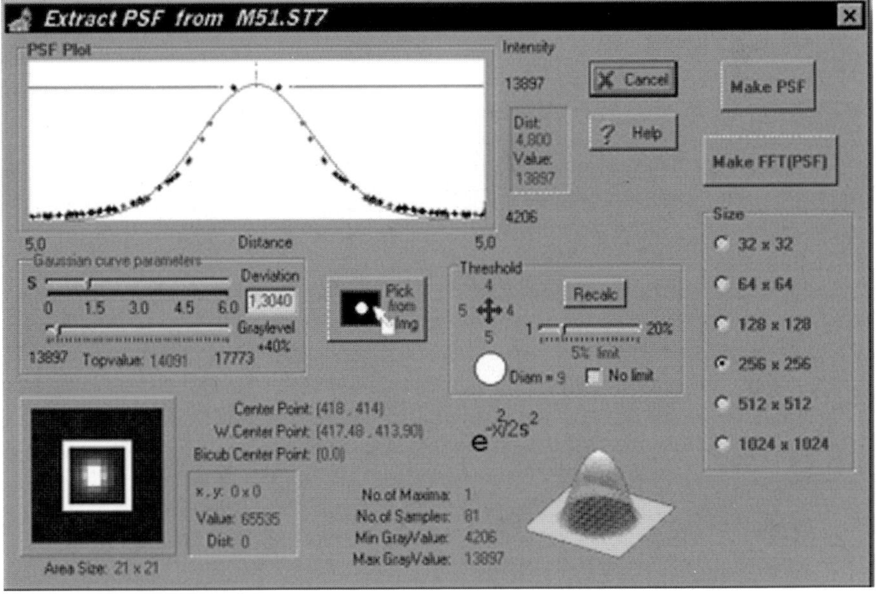

It also has the usual image enhancement commands such as histogram equalisation and an interesting image centroid calculation. It is early days for this promising package and the authors promise more features, including Maximum Entropy Deconvolution, in future versions.

The shareware version is included on the CD-ROM and should enable you to find out, (at least for a 256×256 image) if these exotic techniques will improve your images or whether they are a lost cause. But remember fortune favours the brave!

Paint Shop Pro

Included

on CD-ROM

Having finished our image restoration and/or enhancement, all that is usually required is a few tweaks with a graphics package to ensure it appears just right on the computer screen or on the printout. Paintshop Pro is probably the best shareware graphics package around and despite one unfortunate omission (but see later how to get around it), has all the necessary functions to get the best out of an image, be it on screen or paper. It also makes a good image format converter, although

Figure 8.11. The ubiquitous Paint Shop Pro.

this would be rather to under-use it. One thing I particularly like is the full screen preview where the image is displayed minus surrounding Windows clutter. This is ideal for making 35mm slides direct from the screen. Although it does not centre the image in this mode, this can be overcome by cutting and pasting the image into a larger black one. Minor blemishes in the image can be touched up with a variety of tools and the contrast/brightness adjusted to your personal taste.

Returning to that one important omission I mentioned. It has a sharpen routine but this is crude for astronomical images and seldom results in an improvement. What it really needs is an automatic unsharp mask command. This can be overcome however as follows. First make a copy of the image and then blur or soften it several times. Only trial and error will tell you by how much. Next reduce the brightness by about 50%. Then, using the Image>Arithmetic option, subtract this second image from the original. The result is an unsharp masked new image (Figure 8.12). All that remains is to brighten the resulting image as necessary. In fact doing it manually, one does get a better understanding of what unsharp masking is all about.

Paintshop Pro is shareware and is restricted to 30 days use. It is marketed by JASC Software from whom

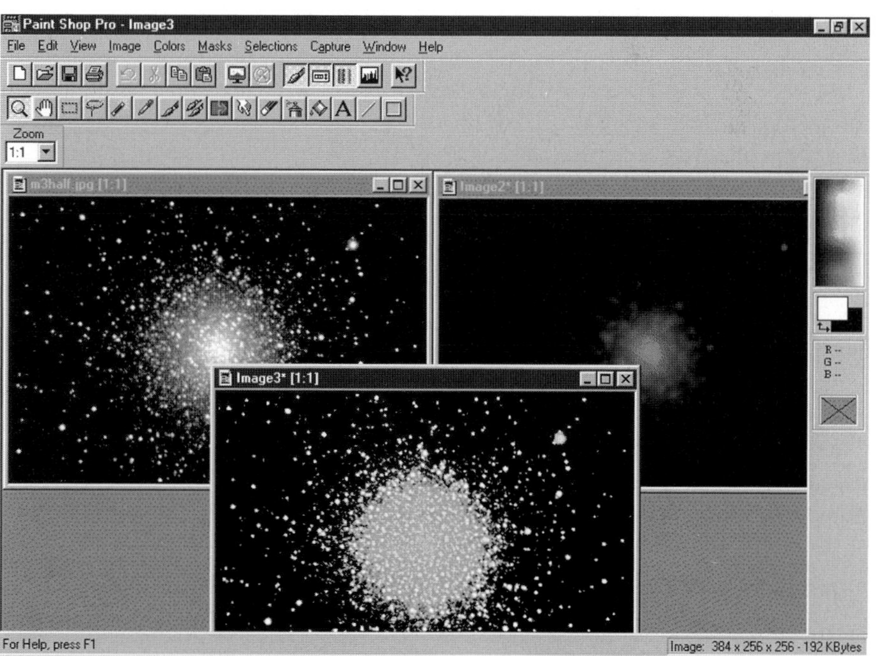

Figure 8.12. Manual unsharp masking using Paint Shop Pro.

the registered version can be purchased. Shareware versions for both Windows 3.1 and Windows 95/98 are included on the CD-ROM in the folder PSP.

Stop Press: Since this chapter was written, version 5 has been released (included on the CD-ROM) and it now has a filter for unsharp masking.

URLs featured in this chapter

The CCD Cookbook Camera Homepage
http://www.wvi.com/~rberry/cookbook.htm

FITSview Homepage
http://www.cv.nrao.edu/~bcotton/fitsview.html

WinMips – no homepage but available from
ftp://ftp.bdl.fr/pub/softwares/pc/mips
Also available here is Qmips32 version 1.3 (freeware)

Quantum Image Homepage
http://www.jamtnet.se/QuantumImage

JASC Software (Paintshop Pro)
http://www.jasc.com

URLs of Amateur CCD Imagers from the Book "The Art and Science of CCD Astronomy"

Adrian Catterall
http://www.observatory.demon.co.uk

Brian Colville
http://www.lindsaycomp.on.ca/~maple

Tim Puckett
http://www.cometwatch.com

David Strange
http://ourworld.compuserve.com/homepages/dstrange

Nik Szymanek and Ian King
http://ourworld.compuserve.com/homepages/Nik_Szymanek

Gregory Terrance
http://www.frontiernet.net/~gregoryt

Luc Vanhoeck
http://ourworld.compuserve.com/homepages/Luc_Vanhoeck

My own web pages (included on the CD-ROM – see chapter 12)
http://ourworld.compuserve.com/homepages/david_ratledge

Chapter 9
Telescope Making

Background

Telescope making is unfortunately not as popular as it once was. Yet there can be nothing to compare with seeing the universe through a telescope crafted with your own hands. The arrival onto the market of computer controlled commercial telescopes, appealing to today's generation of IT literate amateur astronomers, has probably contributed to the myth that the amateur telescope maker can no longer compete. Nothing could be further from the truth. Not only can today's telescope maker compete, he (or she) can and is producing telescopes and control systems exceeding the quality of those from commercial manufacturers. Remember that a mass produced telescope is built to tight budget limits and when it comes to the time to add to it, ancillary equipment or secondary telescopes, those limits could well be a problem. Home made telescopes can be designed to our own exact specification and for our particular requirements. They can, and often now do, include the latest electronic wizardry, as you will see in this chapter. Bells and whistles are not the preserve of the commercial market!

Probably the most common reason for building your own telescope is to obtain a telescope bigger than would otherwise be possible. There is not an amateur astronomer alive who does not dream of a bigger telescope. Thanks to the Internet, these modern day telescope makers can be reached wherever they are in the world. Making a big telescope can be a daunting task on your own, but when you can see and contact amateurs just like yourself, who have already built that dream, the task becomes much less frightening. We can learn from their experiences and expertise to shortcut the learning curve to a new era of amateur telescope making. There really is no excuse for not joining them and you will be following in a great tradition. Lord Rosse and William Herschel were amateur telescope makers after all!

Stephen Tonkin's UK ATM Resources Homepage

What better place to start than the excellent resource centre for UK amateur telescope making run by Stephen Tonkin. Here you will find software to aid the telescope builder and a list of UK suppliers for those hard to find components and supplies (Figure 9.2, *opposite*). Links to suppliers are included where they have a web site of their own.

Figure 9.1. The UK Amateur Telescope Making Resources Homepages.

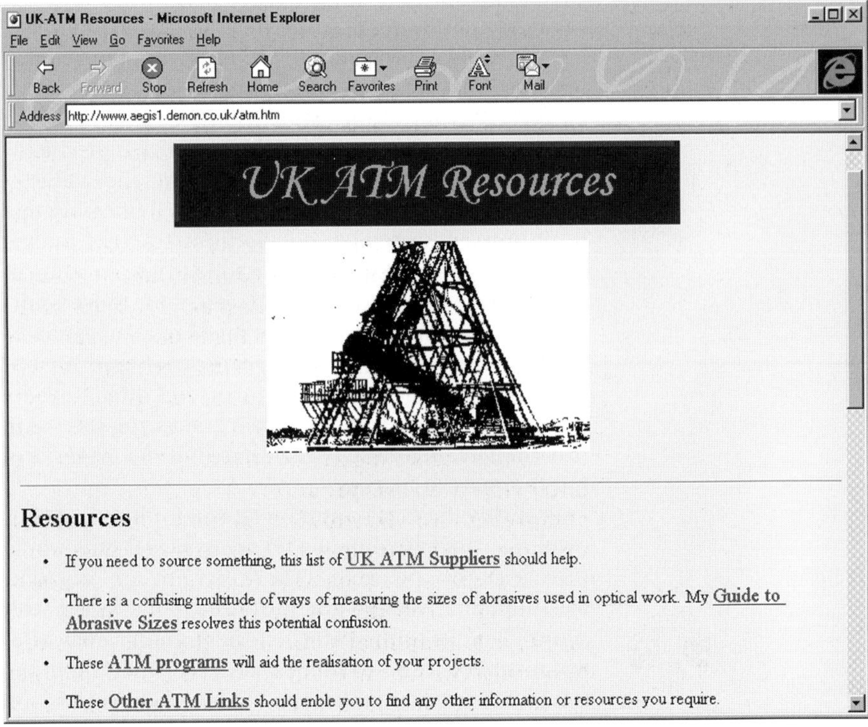

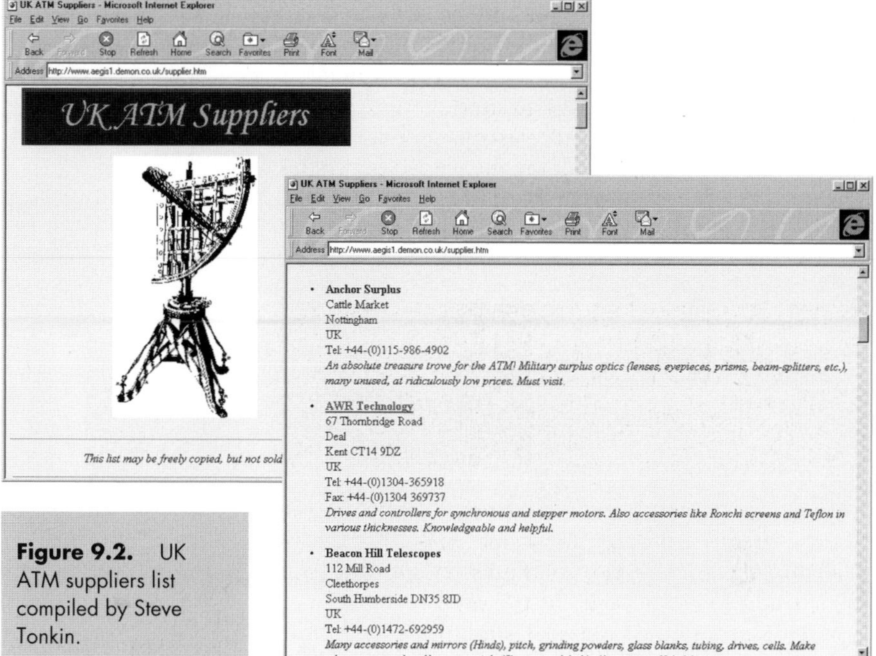

Figure 9.2. UK ATM suppliers list compiled by Steve Tonkin.

Included on CD-ROM

This list of suppliers has been included on the CD-ROM in HTML format so that it can be used for "jumping-off" to their own web pages.

The software on the web site is drawn from a variety of sources and is a mixture of shareware and freeware. Useful utilities include David Chandler's flotation cell design program. This is essential if you are contemplating a large thin mirror as a 9, 18, or even 27 point cell will be required to support it. There are several foucault test result analysers for mirror makers by Jim Burrows, Larry Phillips and Dick Suiter. On the ray tracing front we have one program for achromatic doublets by the team of Jean Prideaux, Donald and Jerry Wright. A second ray tracing program, IRT by Webster Cash, is a more general one for a variety of optics. Gerald Pearson's program deals with Newtonian vignetting as does Newt by Dale Keler. The latter is a slick CAD program which is, in addition, a comprehensive Newtonian design and analysis package. This Windows (3.1 or higher) program is sort of shareware. Dale asks for $10 if you decide to use it but if you do not he invites you to enjoy it anyway! It graphically displays the side view of a Newtonian with or without a superimposed enlarged section of the diagonal/focuser area. It will analyse the

Included on CD-ROM

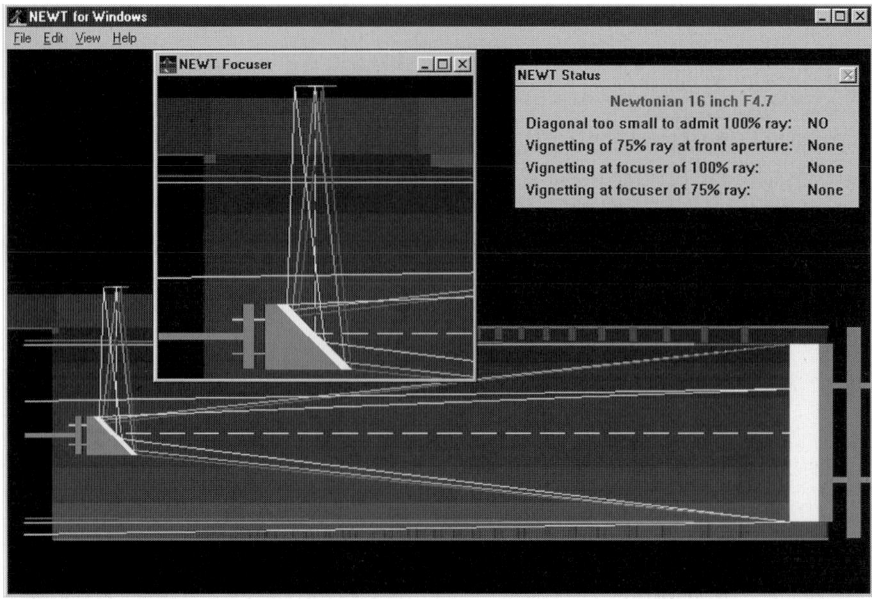

Figure 9.3. The excellent program NEWT, with an enlarged view of the diagonal/focuser area.

effect of different component sizes, including the diagonal mirror, for vignetting calculating fields of view with 100% and 75% illumination. You can even input various eyepieces and check their suitability. Finally you can print out specifications and dimensions. There is a manual in Microsoft Word format. An excellent program which every Newtonian telescope owner, old or new, should have and well worth the price asked.

Another useful source of information on the web site is the ATM Frequently Asked Questions (FAQ). This is maintained by Tim Poulsen. It is in 7 parts and is essential reading for those wishing to make contact with other telescope makers on the Internet. This document is included on the accompanying CD-ROM.

Included

on CD-ROM

Mel Bartels' Homepage

This one of my favourite sites and one I return to regularly. There is almost certain to be something new here for the telescope builder. Mel Bartels is an expert in

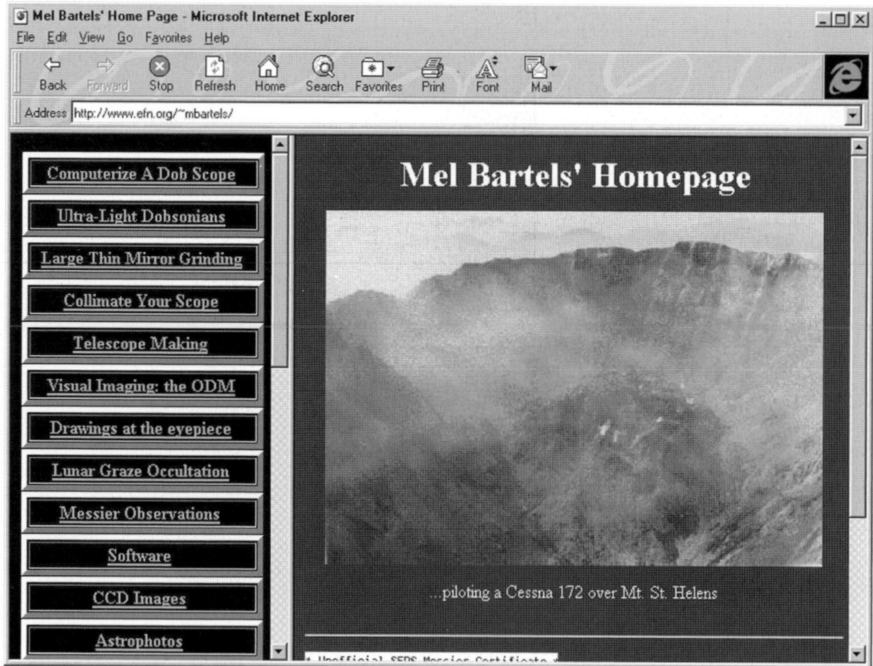

Figure 9.4. Mel Bartels' Homepage.

ultra-lightweight Dobsonian telescopes and through his web pages shares his knowledge and expertise. He is also a pilot, hence the Homepage with a picture taken flying over the Mount St. Helens volcano. Nothing really to do with telescope making but interesting nonetheless (Figure 9.4).

Dobsonian type telescopes are going through a second generation revolution whereby large apertures, when made ultra-light, become easily portable. Designed to collapse down, they can be transported by car to that out of town dark sky. The secret is a low centre of gravity achieved by having a truss type tube and a fast (short focal ratio) mirror. Mel's 20 inch f/5 Dobsonian weighs less than 100 lbs (plus the mirror at 50 lbs) and fits in a small (by USA standards) car. He also explains the grinding, polishing and figuring techniques for making big thin mirrors which are so essential for these telescopes. Inspiration for all of us who dream of a big telescope! Mel's ultra-light Dobsonian description, complete with detailed CAD drawing, is included on the CD-ROM (ul-dob.html).

Included on CD-ROM

However, as if that were not enough, Mel tops it all by providing full constructional details for motorising your telescope with computer control!

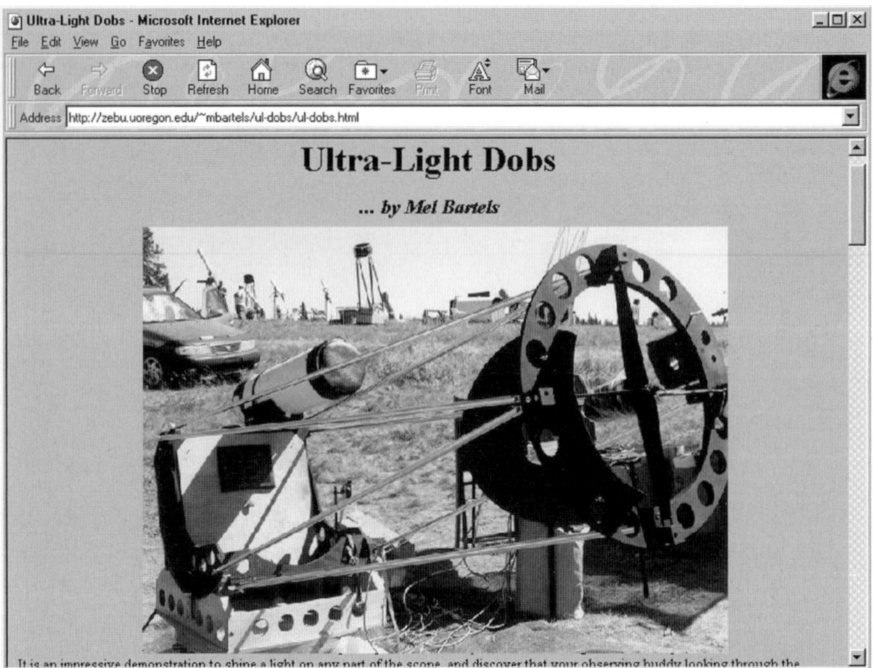

Figure 9.5. Mel Bartels' guide to Ultra-Light Dobsonians.

The field of view of big telescopes is very small so computer controlling it, to enable objects to be located without hassle, is a big step forward. Having found the object tracking is essential, for without it the object will soon disappear before the next person has a chance to look. Mel's computer motorised drive achieves both these objectives by providing digital positioning and tracking. Because altazimuth mounts suffer from field rotation when tracking, Mel even includes details for an optional field de-rotator for long exposure photography or CCD imaging.

The software supplied (Figure 9.7, *overleaf*) can be linked to a planetarium program (Guide) which then takes control and commands the telescope to slew to any object selected.

Thanks to Mel's generosity full details, working plans, circuit diagrams, are all included on the CD-ROM in HTML format (altaz.html). Most components required can be purchased cheaply (even the PC need only be a 386) and several other amateurs have now successfully computerised their own telescope following Mel's instructions. All that Mel asks is a $25 contribution should you decide to follow suit. Why not join the big telescope revolution and with computer control!

Motorize Your Telescope - top page - Microsoft Internet Explorer

File Edit View Go Favorites Help

Back Forward Stop Refresh Home Search Favorites Print Font Mail

Address http://zebu.uoregon.edu/~mbartels/altaz/altaz.html

Motorize Your Telescope

and add computer control

...by Mel Bartels, last updated March 11, 1998

• click here for Spanish version

the 20.4" f/4.8 motorized dob...

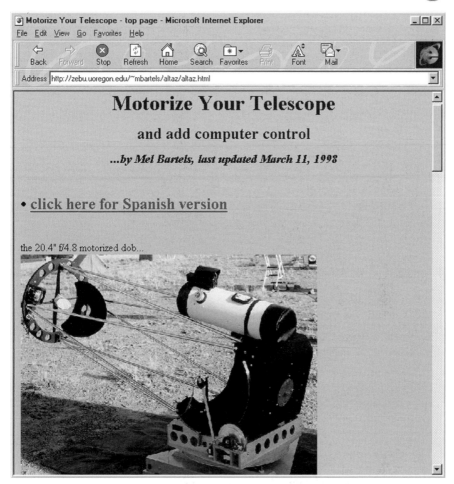

Figure 9.6. Motorise a Dobsonian telescope by Mel Bartels.

Bruce Sayre's 20 inch Telescope

Following on the big telescope theme is Bruce Sayre's highly impressive half metre (20 inch) alt-az telescope. On Bruce's web site there is a variety of telescopes (watch out for a new 14 inch one) but this sophisticated monster stands out. The design and construction are first class (Figure 9.8, *overleaf*).

```
      Altaz Scope Control Program last compiled 12-31-97  by Mel Bartels
Quit                  Guide.bat             Reset to Equat        Move to Altaz
Data File             Handpad               Reset to Altaz        Move to Home
Scroll Tour           Track                 Reset to Home         Input Equat
Init 1                PEC                   Store Equat #1        Input Altaz
Init 2                Ms Speed              Recover Equat #1      Offset Equat
Init 3                Guide Speed           Store Equat #2        Offset Altaz
Optimize Init         Init Encoders         Recover Equat #2      Drift Equat
DOS shell             Reset to Enc.         Move to Equat         Drift Altaz
 --> quit the program

Handpad Switch: Off                    Track Status:  Off  PEC Status:  Off
Microstep:     300"/sec    Guide:       5"/sec   PWM reps:           LX200:
Slew:  Done
File:                                  Object:
Local Time:  11:16:28  Date:    1/02      Sidereal Time:  16:53:00
Refraction:  0.026                        Field Rotate:  299.687
Encoders:Alt:    N/A      Az:    N/A      Counts:Alt:  N/A    Az:  N/A
Current: Alt:    34.875   Az:    52.894   Ra:    7:17:59    Dec:  +49:05:27
Input:   Alt:     0.000   Az:     0.000   Ra:    0:00:00    Dec:  + 0:00:00
Drift:   Alt:     0.000/m Az:     0.000/m Ra:    0:00:00/h Dec:  + 0:00:00/h
Init1: Alt:    7.900 Az:   45.600 Ra:  16:41:42 Dec:  +36:28:13 ST:  0:00:00
Init2: Alt:  -33.000 Az:  347.700 Ra:  21:30:00 Dec:  +12:10:00 ST:  0:00:00
Init3: Alt:   33.000 Az:  105.700 Ra:  12:30:55 Dec:  +12:24:27 ST:  0:00:00
Scope: Lat:   44.265 Long: 351.392  AzOff:  0.393   HAOff: - 8:46:25
```

Figure 9.7. The computerised telescope control program Alt-Az by Mel Bartels.

I can do nothing better than repeat Bruce's own summary of the telescope characteristics:–

- When dissembled for travel, the heaviest component is 45 pounds: the mirror in its cell.
- All parts are waterproof and machined from aluminium. No wood is used.
- The tube moves on two 24-inch diameter C-shaped split rings bolted to the bottom end ring. These split rings rotate on ball bearings in forks mounted on the azimuth platform. The tube swings an unrestricted 180°.
- The upper tube end is a single light ring with a focuser shelf welded to it.
- There is no mirror box. The removable cell rests in the open on the tube's bottom ring.
- Azimuth motion is provided by an open, 26-inch diameter thrust ball bearing mounted on a rigid hexagonal frame. Thumbscrews level the frame.
- Stepper motors, controlled by a Tech2000 Dob Driver II, drive the telescope through spring-loaded friction rollers.
- The maximum eyepiece height is only 6′ 6″. The bottom ring comes within $\frac{1}{2}$″ of the ground as the tube swings through the altitude axis.
- Foam-filled struts are bolted to the top and bottom rings and aligned with spherical washers.

Full details are included on the CD-ROM for this inspirational telescope.

20-inch telescope article - Microsoft Internet Explorer

File Edit View Go Favorites Help

Back Forward Stop Refresh Home Search Favorites Print For

Address http://www.foothill.net/~sayre/20-inch Telescope.htm

20-inch telescope

A front view of the telescope shows its open structure

20" f4 split ring alt-azimuth

Front view of telescope

Table of contents

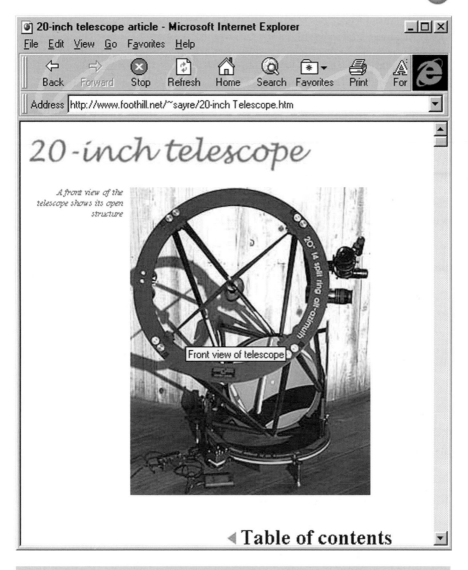

Figure 9.8. Bruce Sayre's magnificent 20 inch split ring mounted telescope.

The Bolton Group Telescope Making

This web site features some of the telescopes made by the Bolton Astronomical Society trio of Gerald Bramall, Brian Webber and myself. Gerald is a self taught machinist specialising in aluminium fabrications, with several stunning telescopes to his name. Brian is an optic maker of some renown having made tens of mirrors, a 5 inch object glass, a Maksutov and a Springfield mounted Newtonian. Myself, I have been making telescopes for 35 years but still suffer from "aperture fever"!

The web pages cover an ever increasing range of telescopes and accessories from a simple 20 cm (8 inch) Dobsonian, for which drawings and a cutting plan for parts is given, to an observatory class 40 cm (16 inch) Newtonian. Probably pride of place however goes to the 8 inch binoculars, which perform as good as they look. They hold collimation perfectly all night, which can be a major problem with Newtonian binoculars. When in use on an open night there is always a long queue to look through them and experience the wonders of the night sky with two eyes. Equally impressive, and always a show stopper with the public, is a 12.7 cm (5inch) classical refractor.

Also featured are several gadgets which make observing more enjoyable, including the group's variation on

Figure 9.9. Just some of the telescopes featured on the Bolton Group web site.

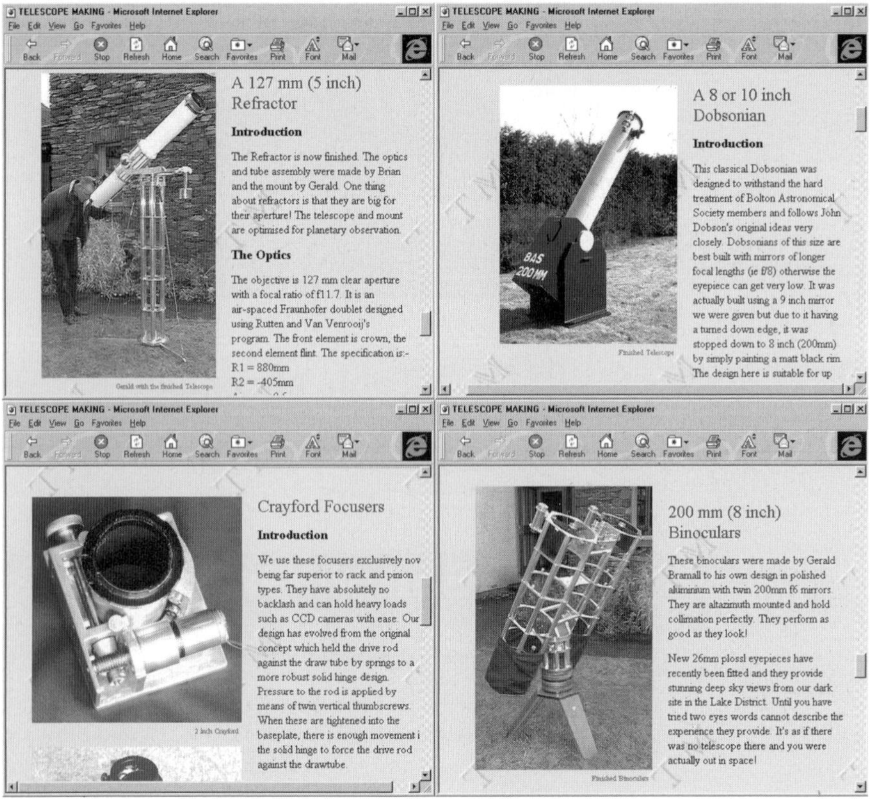

the Crayford type focuser and a device for centring guide stars without moving the telescope. The latest development is in computer control and whilst it is early days in this project the philosophy and initial progress is discussed. These pages are included on the CD-ROM and we welcome feedback.

Amateur Telescope Making Journal

This publication took over when the highly regarded *Telescope Making* magazine, edited by Richard Berry, ceased publication. This new magazine, perhaps newsletter would be a better description, took its place and has its own web site (Figure 9.10). Here you can browse the contents page of each back issue. There is a wide variety of articles with some of the best known telescope makers and optical experts as regular contributors. Some selected articles are published online and can be downloaded just to whet your appetite!

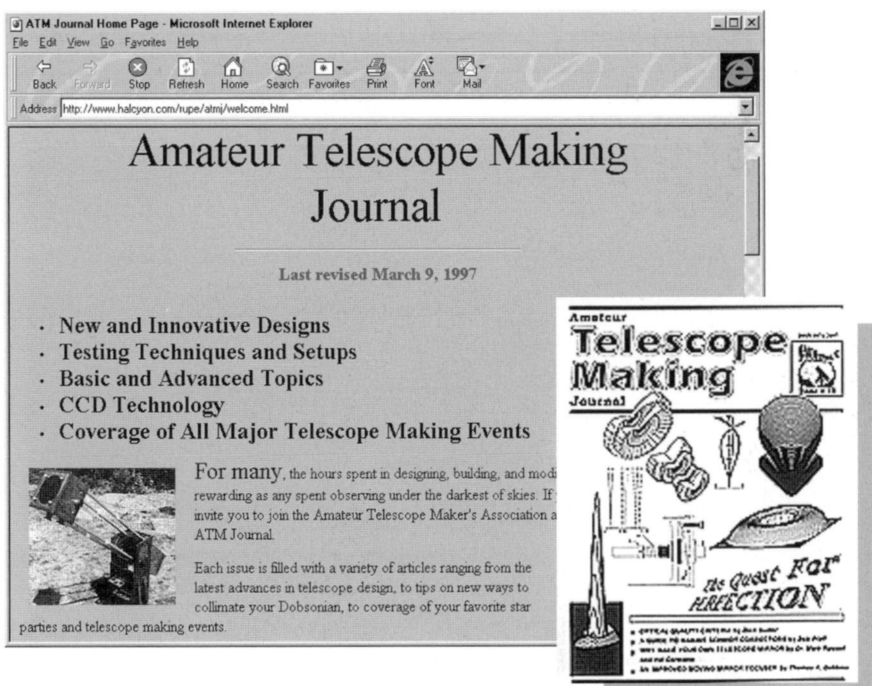

Figure 9.10. The Amateur Telescope Making Journal Homepage.

Overall a site and newsletter worth your support and full subscription details are available at their web site. The URL is given at the end of this chapter.

Tpoint Homepage

Have you ever wondered how professionals get their telescopes to point with arcsecond accuracy? Well many of them use Tpoint software written by the guru of telescope pointing, Patrick Wallace. This British software is a world leader and is compatible with some amateur/professional control systems, such as those by Software Bisque and Comsoft.

The software works by analysing the digital readouts from the telescope when it points at a standard set of stars. By comparing actual positions with apparent readings, Tpoint can work out what is wrong with the telescope, e.g. mount/tube flexure or polar misalignment, and produce a correcting model file. This

Figure 9.11. Patrick Wallace's Tpoint Homepage.

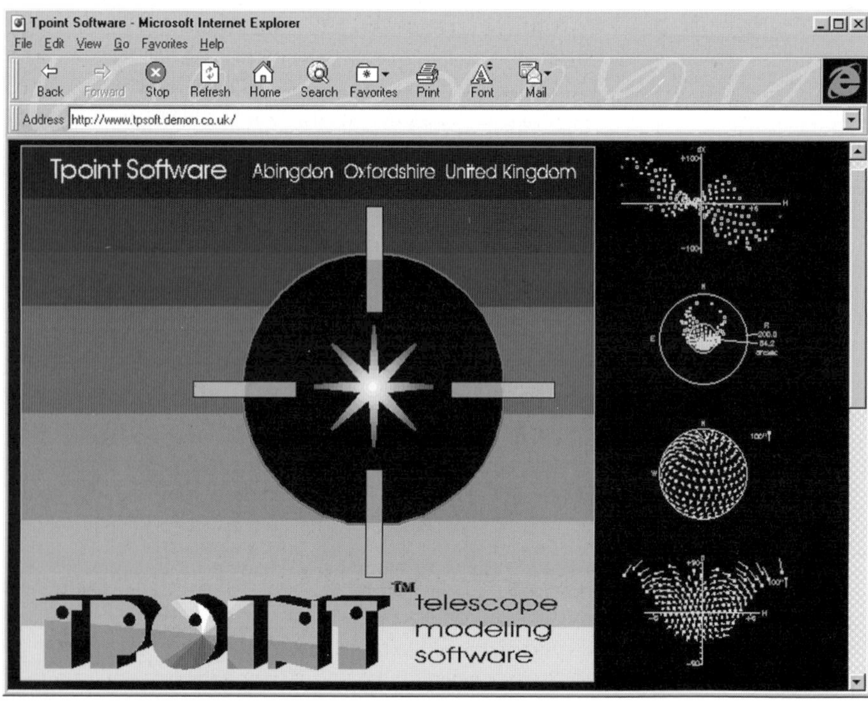

can then be incorporated into the telescope's control system where all these errors can be corrected. On Patrick's web site is an excellent description of how Tpoint was applied to some of the world's largest telescopes, the Palomar 200 inch and the Anglo Australian Telescope and the dramatic pointing accuracy improvement which resulted. An amateur installation is also discussed. Patrick has kindly agreed to his article being included on the accompanying CD-ROM. If you are considering computer control then Tpoint could be just what you are looking for. Even if you are not, the article is recommended reading for the insight it provides.

Comsoft Telescope Control System (PC-TCS)

If you already have a telescope and are looking to add computer control, then one professional system, compatible with Tpoint and affordable to amateurs, is Comsoft's PC-TCS. This sophisticated system can control all manner of telescope functions including, slewing, tracking (at various rates), focusing, mirror covering and even dome control. Unlike most other systems, to keep costs down, it is designed to be run on an old DOS PC, a 386 or 486 is fine. Written by Dave Harvey, who will supply either complete systems, i.e. software and hardware, or a kit of parts for do-it-yourselfers. PC-TCS is in use on a variety of both professional and amateur telescopes around the world, including one at the South Pole!

Included on CD-ROM

A demo version of PC-TCS is included on the CD-ROM and whilst the database is limited to just a few objects on to which you can slew and track, it does give an interesting indication of sophisticated control systems (see Figure 9.12, *overleaf*).

URLs Featured in this Chapter

Stephen Tonkin's ATM Resource Centre
http://www.aegis1.demon.co.uk/atm.htm

```
UT DATE: 07/02/1995        UT: 04:01:46      LST: 15:17:46      LHA: -01:03:25.62
     JD: 2449900.67       UTC: Disabled                              Tube EAST
 ACCESS: NOVICE            RA                        DEC
                                                                El:      34.9
AR_ Epoch: J2000.0    16:21:27.29    TRACK+ro    -20:46:00.3    Az:    +161.8
                                                                Secz:    1.75
    Commanded:        16:21:27.31   SLEW PERMIT  -20:46:00.3    PA:     -16.4

    Next:             16:21:27.31    Jupiter     -20:46:00.3    Mag:    -2.50
    Reference:        15:14:44.84                +32:13:21.7
    Offset Vector:     :  :  .                    :  :  .
δ   Wobble Vector:     :  :  .                    :  :  .       IIS:      128
δ   Difference:              +0.0                       +0.0    Foc:   -2.000
                                                                Dome:   +71.8
    Bias:                 +0.0000                   +0.0000
δ   Guide:                +3.0000                   +3.0000
δ   Drift:               +40.0000                  +40.0000

  Coordinate Mode:  File                    Command File: AUTOSTOP.TCS
  Coordinate File:  SAO.PCF              Current Command: WAITNONEXT
 Coordinate Epoch:  TOD       MPNAR_p      Command Count:   0

                        Runtime Operations Menu
Motion  deClare  mOve  Position  Rates  Source  Time  pArameters  shutDown
```

Figure 9.12. Mission control for PC-TCS.

Mel Bartels' Homepage
http://www.efn.org/~mbartels

Bruce Sayre's Homepage
http://www.foothill.net/~sayre

The Bolton Group – Telescope Making
http://ourworld.compuserve.com/homepages/david_ratledge /tm.htm

Amateur Telescope Making Journal
http://www.atmjournal.com

Patrick Wallace's Tpoint Homepage
http://www.tpsoft.demon.co.uk

Comsoft PC-TCS Control Systems
http://www.comsoft-telescope.com

Useful URLs not Referred to but Recommended

The ATM Page – this USA web site is similar to Steve Tonkin's but is a bit bigger and has an American emphasis. It would have been included in the chapter but was down for 3 months at the time of compiling this book. Hopefully it will have re-appeared.
http://atmpage.com

Manchester Astronomical Society
http://www.u-net.com/ph/mas/home.htm

Tallahasse Astronomical Society
http://www.polaris.net/~tas/atm/at..htm

USA ATM Resource List
http://www.freenet.tlh.fl.us/~blombard

Chapter 10

Satellite Tracking

Background

Satellite tracking is popular. You only have to see the counts of the number of people downloading satellite orbital elements to realise this. By the end of the day on which new elements are published, thousands will have logged on and copied them into their satellite tracking program. I can remember as a boy seeing one of the early Sputniks going overhead – it was visible proof that the USSR had taken an early lead in the space race. From the early days through to those of the MIR space-station today, satellite observation has always been of great public interest. In this chapter we look at just three of the many tracking programs available, which predict what will be passing over your particular observing site tonight. If you are new to satellite tracking software, one thing to be aware of is that orbital elements for satellites go out of date very rapidly. If you get hooked on satellite tracking then Internet access will be essential for downloading the latest elements. There are two sections on sites where elements can be obtained but just a word about the jargon. Satellite observing, like most specialist areas, has its own "techno speak". Elements are sometimes referred to as Keps, short for Keplerian Elements, whilst TLE, is short for Two Line Elements and Elsets, is short for element sets.

For those perhaps not interested in doing it themselves we first look at the Marshall Space Flight Center (their spelling) where you can use their online applications to check satellite visibility. Whichever way, whether you do-it-yourself or have it done for you, there should be something here for you.

Marshall Space Flight Center Liftoff

NASA's Liftoff web site at the Marshall Space Flight Center is a cracker, exploiting modern Internet technology to provide an interactive experience for visitors. Stars of the site are two Java applications J-Pass and J-Track. The applications require version 3 of Internet Explorer or Netscape Navigator to work. Even if you only have occasional interest in orbiting satellites, these pages are well worth bookmarking.

J-Pass was developed by Patrick Meyer and Tim Horvath and is designed to locate satellites passing over your sky. First task is to enter your latitude and longitude manually as it only has USA cities pre-loaded. Then select a satellite from the list, I chose MIR. Select "night pass" and click on "next pass" then the application will

Figure 10.1. NASA's Liftoff – Satellite Tracking Homepage.

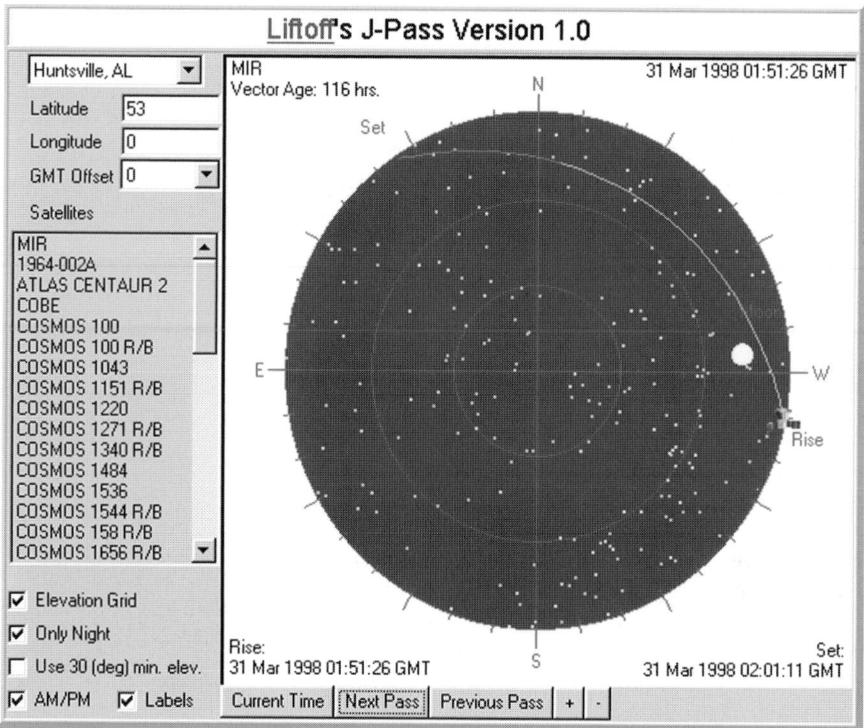

Figure 10.2. Liftoff's J-Pass application showing the path of MIR passing the Moon and Cassiopeia.

calculate the next time MIR (or your selected satellite) will pass overhead. It will be plotted against the stars. In Figure 10.2 you can perhaps make out the Moon with MIR's path crossing Cassiopeia at 1:51a.m.. At present (version 1.0) there is no print facility. You will have to save the screen to the clipboard (Alt+PrtSc) and paste into a graphics package for printing. That apart, J-Pass is very impressive.

J-Track on the other hand is more of a "Mission Control" type application for displaying just where, over the Earth, satellites are located and in real-time. You can even watch them move! Again there is a wide selection of satellites, such as MIR, the Space Shuttle, Hubble Space Telescope, UARS and COBE. We are not limited to displaying just one either, we can have them superimposed one on another. The satellites are shown against a map of the world with the day/night areas shaded accordingly. The position of the Sun is also shown (Figure 10.3, *overleaf*). One interesting option is to have the current weather map draped over the plot (Figure 10.4, *overleaf*). This takes a minute or so while the data is retrieved but is very impressive.

J-Track 2.0

UARS
MIR
HST
30 Mar 1998 19:46:21 GMT
MIR Lat: 37.3 Lon: -4.5 Alt: 382.9
Config.
Use the "Config." button at the left to change satellites, turn on the weather, etc. Questions? ...More
Next

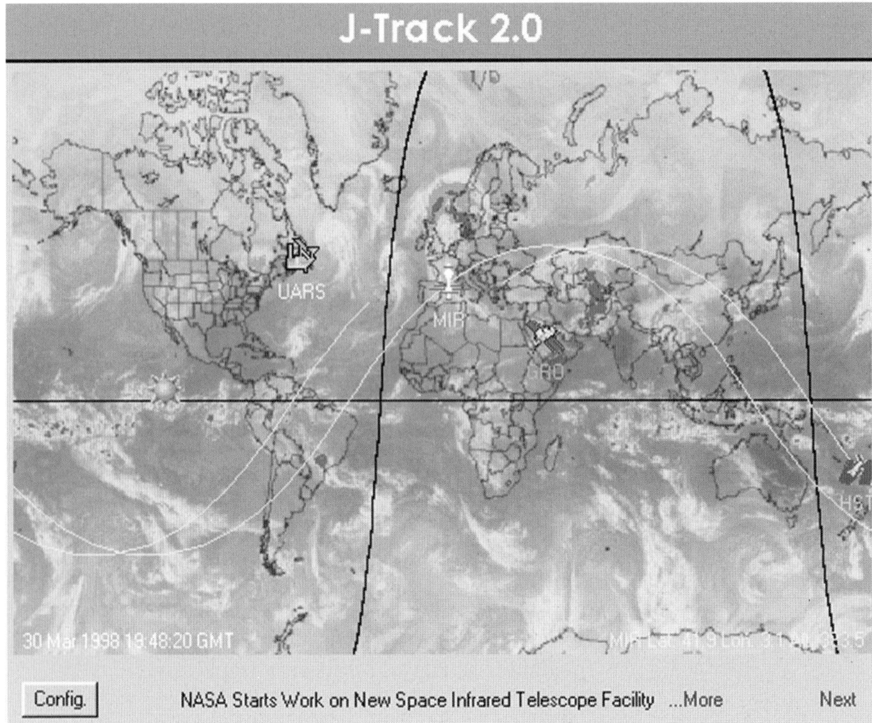

The Liftoff site also contains much other information but it will be J-Pass and J-Track which will draw you back to this site over and over again.

SATTRACK Homepage

The SATTRACK Satellite Tracking web site is a good place to start if you are new to this specialist field. It has the latest Keplerian Element Files (see Figure 10.6, *overleaf*). These are updated frequently – sometimes as often as every day. It also has much useful information for the beginner with an explanation of the element formats and a satellite spotters guide. Some of the many tracking programs are available for downloading.

◄
Figure 10.3.
Liftoff's J-Track showing MIR and the Hubble Space Telescope.

CelesTrak Homepage

CelesTrak, run by Dr TS Kelso, is very similar to the Sattrack site. However, the Keplerian elements provided are more logically arranged (at least they seem so to a beginner like me) with them split into sections for:–

- special interest satellites e.g. MIR and the 100 brightest ones
- weather/Earth resources satellites
- communication satellites including geostationary ones and the new Iridium type
- navigational satellites

If you are wondering what Keplerian elements look like then here is a sample in two line format:

```
Mir
1 16609U 86017A   98086.17064438  .00010049  00000-0  11398-3 0  3654
2 16609   51.6578 101.4594 0005038 157.1475 202.9737 15.63049514691224
SPOT 1
1 16613U 86019A   98084.42637168 -.00000044  00000-0  00000+0 0  1773
2 16613   98.7573 159.5836 0001166 153.3286 206.7955 14.20018429312020
Cosmos 1766
1 16881U 86055A   98085.55603874  .00000830  00000-0  96541-4 0  5486
2 16881   82.5176 283.8890 0019790   6.0853 354.0589 14.84034843629843
EGP
1 16908U 86061A   98085.25745298 -.00000083  00000-0  10000-3 0  2547
2 16908   50.0123 167.6300 0011464 220.5723 139.4258 12.44418500196970
NOAA 10
1 16969U 86073A   98086.13123801  .00000140  00000-0  78253-4 0  4411
2 16969   98.5680  77.0984 0012056 264.7878  95.1924 14.25092180598880
MOS-1
1 17527U 87018A   98084.16035203 -.00000043  00000-0  00000+0 0  6573
2 17527   98.8556 139.9257 0017492 338.3206  21.7247 14.00506904565401
```

◄
Figure 10.4.
Liftoff's J-Track with weather data overlaid.

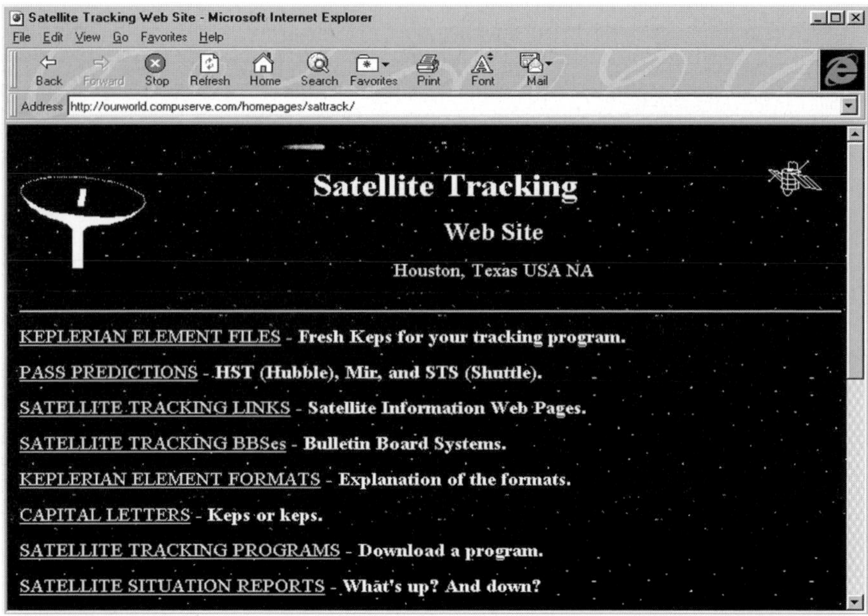

Figure 10.5. The Sattrack Satellite Tracking Web Site.

Figure 10.6. Sattrack download page for the latest Keplerian elements.

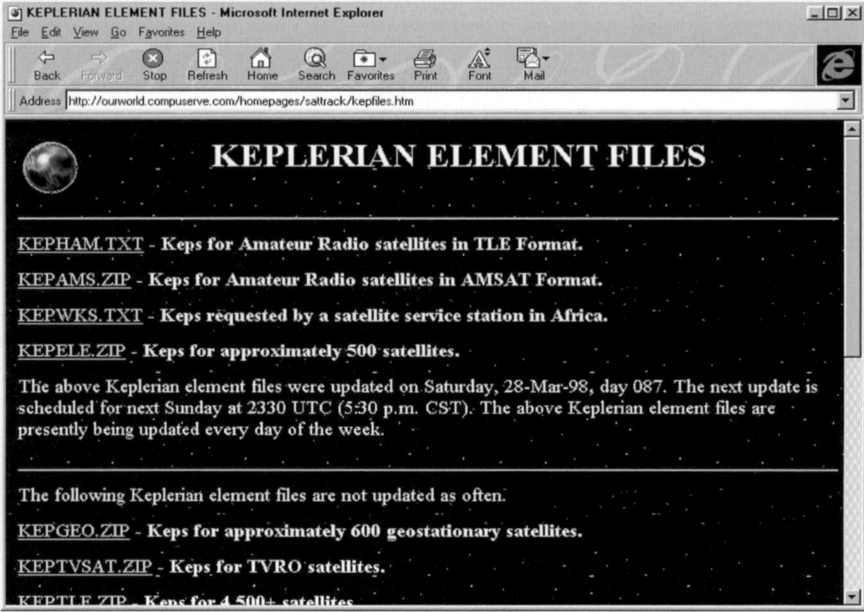

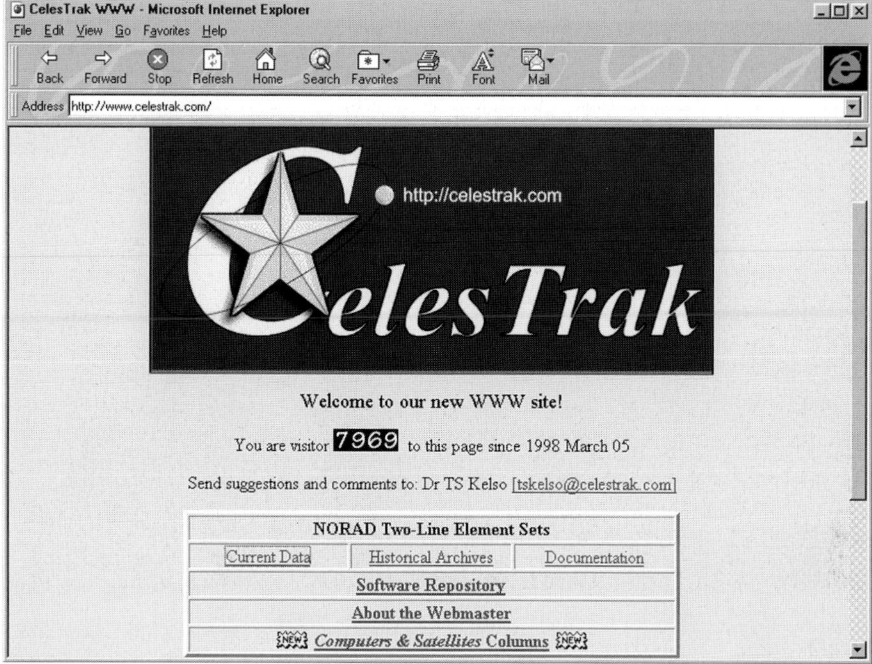

Figure 10.7. CelesTrak Homepage.

TRAKSAT for DOS

Written by Paul Traufler, Wintrak and TRAKSAT are general purpose satellite tracking programs for Windows and DOS respectively. Wintrak is a purchase only product but Traksat is available as shareware and is included on the CD-ROM for you to try along with a sample database. Do not be put off by it being a DOS program. Version 4 has its own graphical interface and is very impressive. It even featured in the James Bond movie, Goldeneye!

TRAKSAT generates predictions derived from the two line elements. It has an extensive list of observing stations, including many towns in the UK. Once having loaded the elements and selected the observing station, Manchester in my case, it will rapidly give you a list of those satellites currently visible. It has a multi-tracking facility whereby up to 6 satellites can be simultaneously tracked and displayed. Output options are many with several text printouts but it is the graphical displays which catch the eye. Satellites can be shown against the

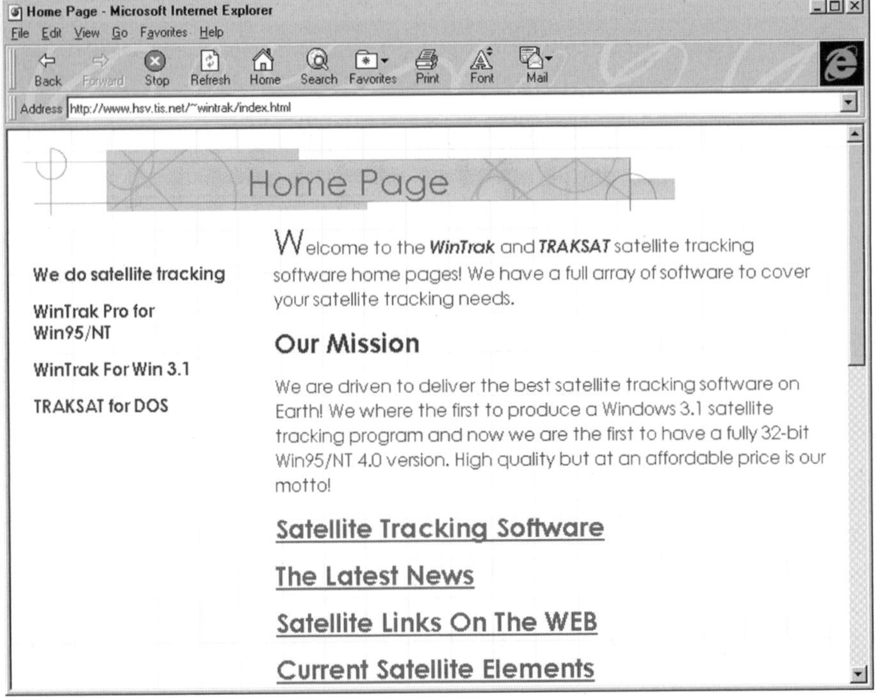

Home Page - Microsoft Internet Explorer
File Edit View Go Favorites Help
Back Forward Stop Refresh Home Search Favorites Print Font Mail
Address http://www.hsv.tis.net/~wintrak/index.html
Home Page
We do satellite tracking
WinTrak Pro for Win95/NT
WinTrak For Win 3.1
TRAKSAT for DOS
Welcome to the WinTrak and TRAKSAT satellite tracking software home pages! We have a full array of software to cover your satellite tracking needs.
Our Mission
We are driven to deliver the best satellite tracking software on Earth! We where the first to produce a Windows 3.1 satellite tracking program and now we are the first to have a fully 32-bit Win95/NT 4.0 version. High quality but at an affordable price is our motto!
Satellite Tracking Software
The Latest News
Satellite Links On The WEB
Current Satellite Elements

Figure 10.8. Homepage for Wintrak and TRAKSAT.

background stars or looking down on Earth. Figure 10.9 (*opposite*) is just one of several different ways of depicting orbits and visibility.

To run TRAKSAT, I would advise restarting your computer in MS-DOS mode, to free up as much memory as possible. A comprehensive 70 page manual is included and should be read before starting. An excellent program well worth the $25 registration fee asked.

SatSpy for Windows

Included

on CD-ROM

This shareware package for Windows is one of the most popular and it is easy to see why when you run it. This version (2.5) is shareware and is limited to 30 days use, ample time for you to be able to learn its many features. Written by David Cappellucci, it has a highly graphical interface with the usual Windows pull-down menus and icons to make life easier.

It has no less that 3 orbit propagators (routines that calculate passes) for a variety of needs and has support for geosynchronous satellites. There are a wide variety

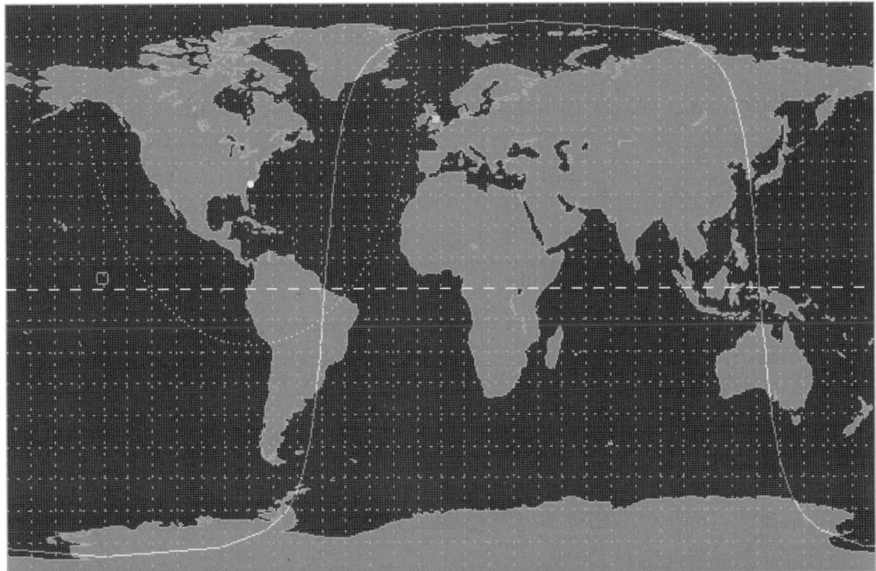

Figure 10.9. TRAKSAT display for MIR over Manchester.

of display modes including plots against the night sky with stars, constellations, bright stars and planets shown, making identification easy. In addition, there are both 3D and flat map plots with the circle of visibility of the selected satellite overlaid. It also shows whether the satellite will be sunlit or in shadow, a fundamental factor in its visibility.

Figure 10.10. The SatSpy Tracking Program.

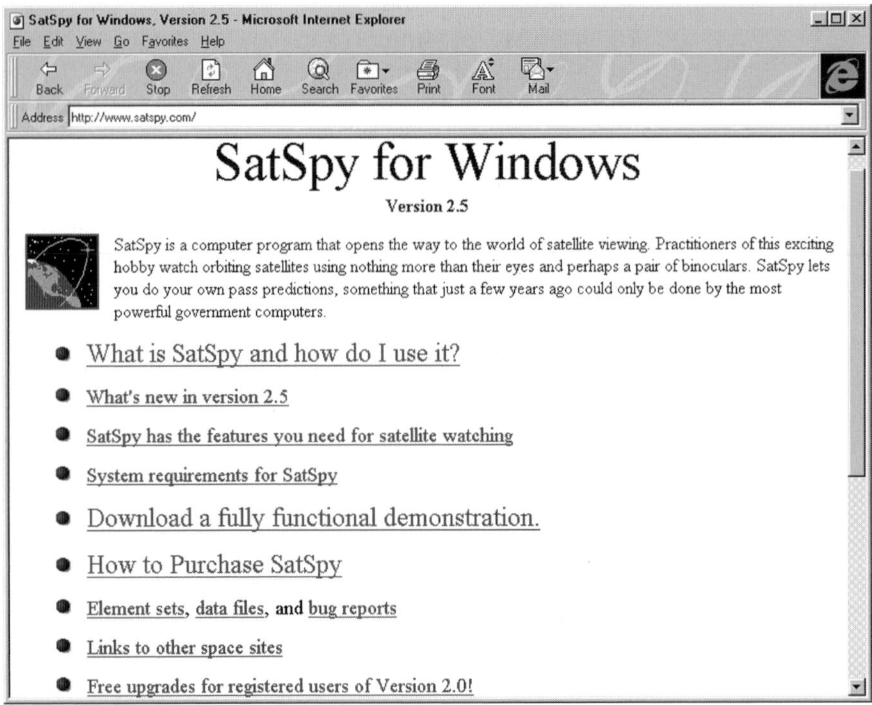

SatSpy for Windows, Version 2.5 - Microsoft Internet Explorer
File Edit View Go Favorites Help
Back Forward Stop Refresh Home Search Favorites Print Font Mail
Address http://www.satspy.com/
SatSpy for Windows
Version 2.5
SatSpy is a computer program that opens the way to the world of satellite viewing. Practitioners of this exciting hobby watch orbiting satellites using nothing more than their eyes and perhaps a pair of binoculars. SatSpy lets you do your own pass predictions, something that just a few years ago could only be done by the most powerful government computers.
What is SatSpy and how do I use it?
What's new in version 2.5
SatSpy has the features you need for satellite watching
System requirements for SatSpy
Download a fully functional demonstration.
How to Purchase SatSpy
Element sets, data files, and bug reports
Links to other space sites
Free upgrades for registered users of Version 2.0!

Figure 10.11. Homepage for SatSpy for Windows.

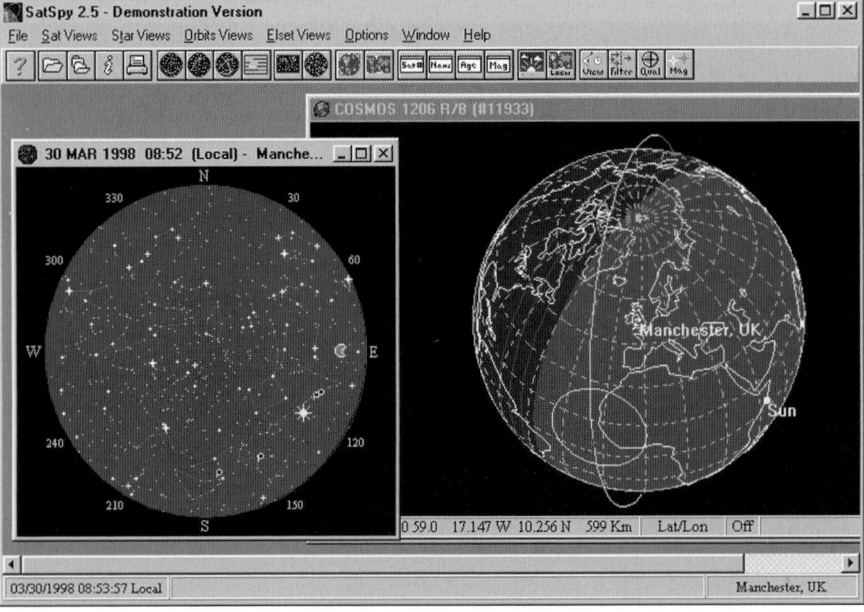

Whilst few non USA viewing locations are included it is an easy matter to enter your own and importantly, its time zone. There is an excellent help facility and also included are Satellite Viewing Tips, an introductory article by the SatSpy author, David Cappellucci, which is essential reading for the beginner and is included on the CD-ROM. If I had to select one satellite tracking package then I would probably pick SatSpy. It is that good.

Included

on CD-ROM

SATPRO for DOS

The third satellite system is again a DOS program but it has impressive graphical displays nonetheless. The full version can "fly" 200 satellites simultaneously but the demo version, included on the CD-ROM, is limited to 6. Developed by David Harvey of the Steward Observatory of the University of Arizona to support not just observing but the imaging of satellites as well. Early versions of the software were used to obtain photographs of MIR in 1987 and the Hubble Space Telescope in 1991.

Perhaps not as intuitive to use as the other programs, you need to use the arrow keys to move around the menus. It is nonetheless a powerful system with facilities such as ESTES, Earth Satellite Temporal Event Scanner (available in the registered version only). This enables you to enter a time and region of the sky where you saw (or imaged) a satellite and it will search its database (up to 20,000 objects) and try and find a match. To run, I would advise restarting your computer in MS-DOS mode and type "Satpro.exe".

◄ **Figure 10.12.** Just two of the display modes for SatSpy, against the constellations and a 3D orbit view.

URLs Featured in this Chapter

Marshall Space Flight Center – Liftoff
http://liftoff.msfc.nasa.gov
http://liftoff.msfc.nasa.gov/realtime/jpass (J-Pass)
http://liftoff.msfc.nasa.gov/realtime/jtrack (J-Track)

SATTRACK Homepage
http://ourworld.compuserve.com/homepages/sattrack

CelesTrak Homepage
http://www.celestrak.com

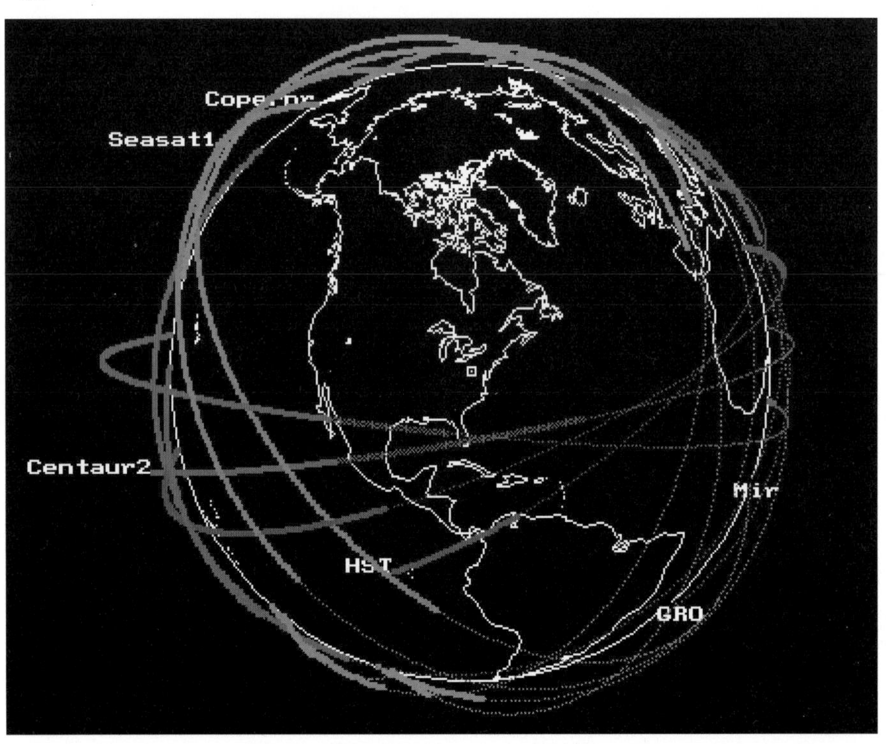
Seasat1
Centaur2
Mir
HST
GRO

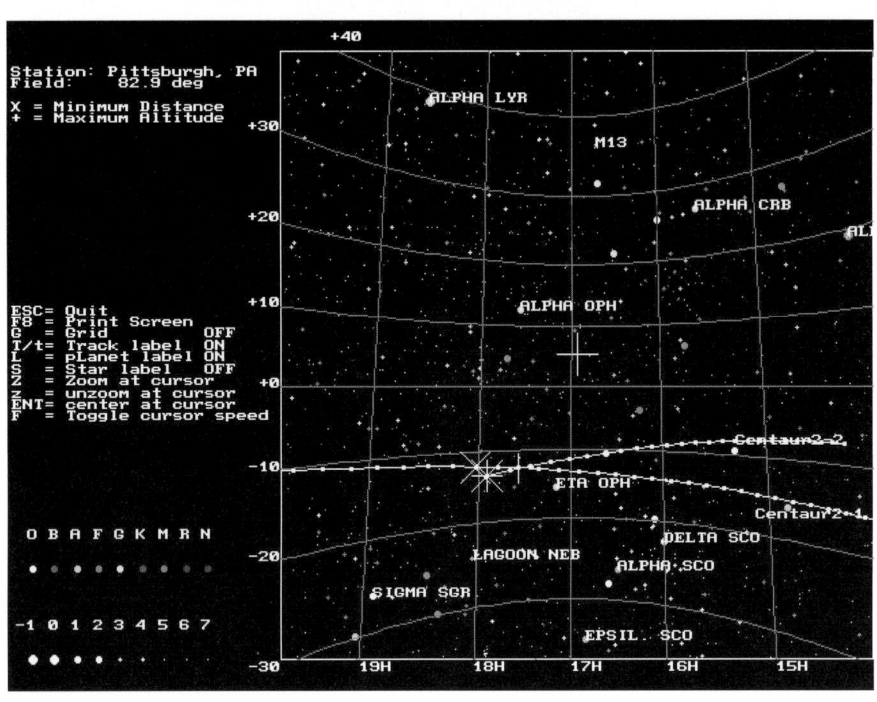
Station: Pittsburgh, PA
Field: 82.9 deg
X = Minimum Distance
+ = Maximum Altitude
ESC= Quit
F8 = Print Screen
G = Grid OFF
T/t= Track label ON
L = pLanet label ON
S = Star label OFF
Z = Zoom at cursor
z = unzoom at cursor
ENT= center at cursor
F = Toggle cursor speed
O B A F G K M R N
-1 0 1 2 3 4 5 6 7
+40
+30
+20
+10
+0
-10
-20
-30
19H
18H
17H
16H
15H
ALPHA LYR
M13
ALPHA CRB
ALPHA OPH
Centaur2-2
ETA OPH
Centaur2-1
DELTA SCO
LAGOON NEB
ALPHA SCO
SIGMA SGR
EPSIL. SCO

TRAKSAT and WinTrak
http://www.hsv.tis.net/~wintrak

SatSpy for Windows
http://www.satspy.com

Comsoft's Satpro
http://www.primenet.com/~comsoft/satpro.htm

◄
Figure 10.13. 3D type plot from SatPro.

Useful URLs not Referred to but Recommended

Orbital Information Group – the definitive source for Elsets for all orbiting bodies
http://oigsysop.atsc.allied.com
It is part of US Space Command
http://spacecom.af.mil/usspace

Visual Satellite Observers – good introduction and has software for downloading
http://www2.satellite.eu.org/sat/vsohp/satintro.html
For information on Iridium Satellites
http://www.satellite.eu.org/sat/vsohp/iridium.html

Spacewarn – latest news
http://nssdc.gsfc.nasa.gov/spacewarn

Satellite Related Software
http://www2.plasma.mpe-garching.mpg.de/sat/vsohp/orbsoft.html

GSOC Satellite Predictions
http://www.gsoc.dlr.de/satvis

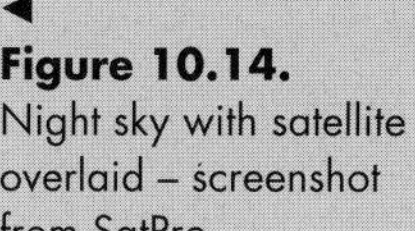

◄
Figure 10.14. Night sky with satellite overlaid – screenshot from SatPro.

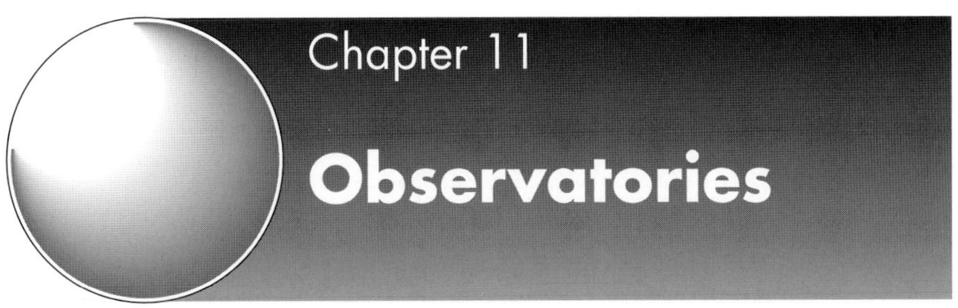

Chapter 11
Observatories

Background

Most of the World's great observatories now welcome visitors. Not necessarily to their actual observatory but to their virtual one on the Internet. In these days of public accountability it makes good sense for them to "open" their doors to the public and provide us with information on their facilities and research. We can visit the great observatories of the past and read about the next generation ones, and all from the comfort of our armchair. Famous ones like Palomar, Yerkes, Mount Wilson, Siding Spring and Jodrell Bank all have extensive Homepages. We will not be able to look through the great telescopes of course, nobody does these days anyway, but we can learn all about them and see their latest images. If you would like to visit them in reality then, where this is permitted, you can find out opening times and directions to locate them.

There are so many observatories with their own Homepages that only a small sample can be included here. My apologies for those I have omitted.

Palomar Observatory

This one had to be first. The California Institute of Technology's Palomar Observatory is a legend in its

Figure 11.1. The legendary 200 inch telescope at the Palomar Observatory.

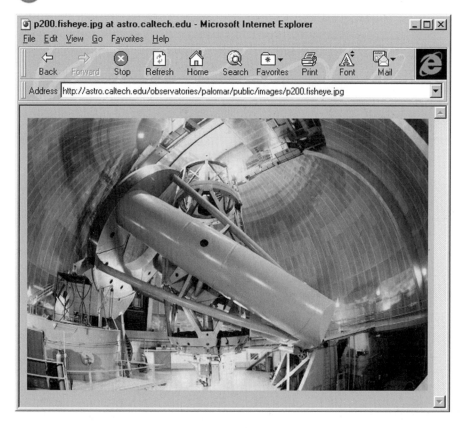

own lifetime. Home to the famous Hale 200 inch – I have never quite got used to it being called the "5 metre". It is still probably the most famous telescope in the world today and many popular books on astronomy over the years have been filled with photographs taken with this telescope. Completed in 1948, after delays caused by World War II, the telescope proved an immense success. It was an ambitious project for its time, being double the size of the world's previous largest, the Mount Wilson 100 inch. Famous controversies, such as that regarding the interpretation of redshifts, were largely resolved with research carried out using its immense light gathering power. The observatory is also home to the 48 inch Schmidt Camera with which the famous sky survey (known as POSS) was carried out. These form much of the Digitised Sky Survey we met in Chapter 7.

The observatory publishes an excellent visitor brochure which is included on the accompanying CD-ROM.

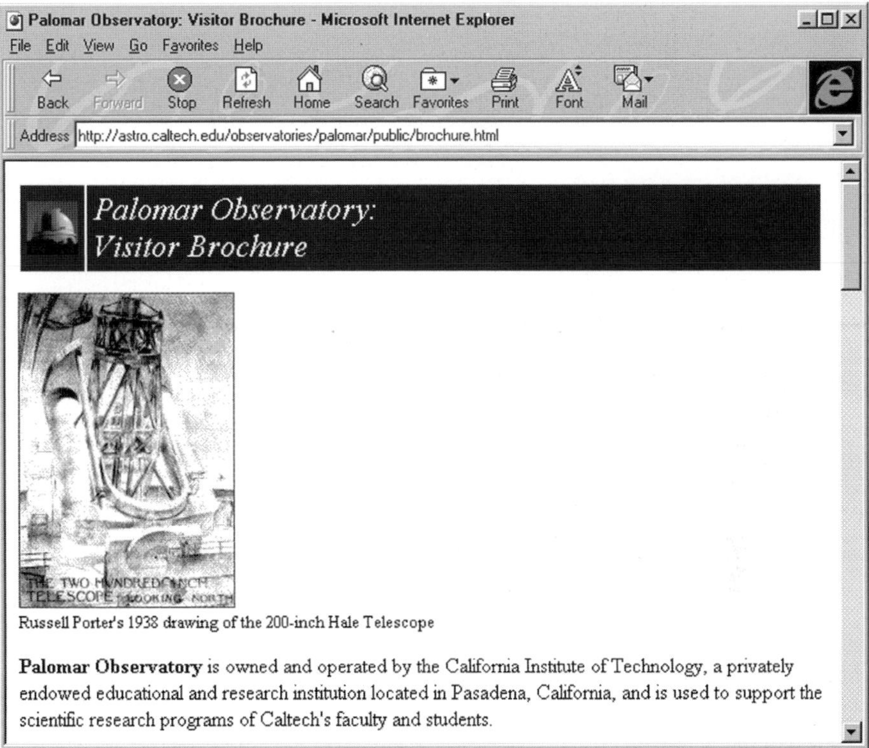
Palomar Observatory: Visitor Brochure - Microsoft Internet Explorer
File Edit View Go Favorites Help
Back Forward Stop Refresh Home Search Favorites Print Font Mail
Address http://astro.caltech.edu/observatories/palomar/public/brochure.html
Palomar Observatory:
Visitor Brochure
Russell Porter's 1938 drawing of the 200-inch Hale Telescope
Palomar Observatory is owned and operated by the California Institute of Technology, a privately endowed educational and research institution located in Pasadena, California, and is used to support the scientific research programs of Caltech's faculty and students.

Figure 11.2. Palomar Observatory's Visitor Brochure (see CD-ROM).

Yerkes Observatory

The University of Chicago's Yerkes Observatory is home to a number of telescopes but it is the great 40 inch refractor, the world's largest, for which it is most famous. The observatory celebrated its centennial year in 1997 and was built in the classical Romanesque style with the 92 feet diameter dome dominating one end of the elevation (Figure 11.3, *overleaf*).

The telescope was the dream of George Ellery Hale (later of Palomar fame) but for funding he had to turn to Charles T. Yerkes, an entrepreneur of dubious methods. Yerkes used his funding of the great telescope to add some respectability to his less philanthropic dealings. Nevertheless Hale got his dream telescope. The great object glass was made by the renowned optician Alvin G. Clark with the tube and mount by Warner and Swasey. Although now dwarfed by huge reflectors, the telescope contributed much to astronomical research and many famous astronomers, including Hubble, Kuiper and Barnard all worked with it.

Yerkes Observatory Virtual Museum - Microsoft Internet Explorer

File Edit View Go Favorites Help

Back Forward Stop Refresh Home Search Favorites Print Font Mail

Address http://astro.uchicago.edu/home/web/yerkes/virtualmuseum/

Yerkes Observatory Virtual Museum

Timeline
People
Instruments
Centennial Events
Credits

Yerkes Observatory Virtual Museum

Welcome to the Yerkes Observatory Virtual Museum. Yerkes Observatory, a department of the University of Chicago, was dedicated to the science of astronomy and astrophysics on October 21, 1897. This is our centennial year. To celebrate, we have created this museum to tell the exciting story of the building of the world's largest refracting telescope and the people whose tireless energy created this beautiful Observatory.

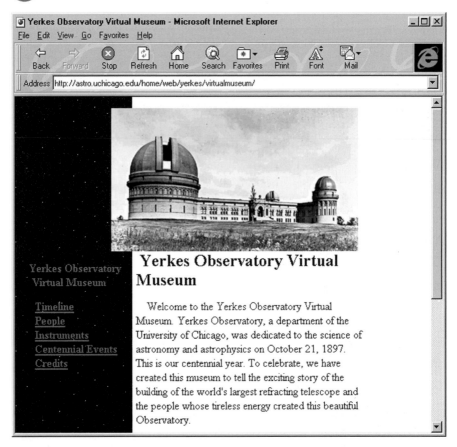

Figure 11.3. The Yerkes Observatory Virtual Tour with the fabulous old buildings in the background.

The web site has a virtual museum tour which includes an informative history of the 40 inch refractor written by Patricia Roussel (Figure 11.4, *opposite*). This is included on the CD-ROM and is essential reading for all those interested in the history of telescopes.

Mount Wilson Observatory

This is another observatory which featured in many of this century's major astronomical discoveries. The

Yerkes Observatory Virtual Museum-Instruments... - Microsoft Internet Explorer

File Edit View Go Favorites Help

Back Forward Stop Refresh Home Search Favorites Print Font Mail

Address http://astro.uchicago.edu/home/web/yerkes/virtualmuseum/40inch.html

40-Inch Refracting Telescope

Yerkes Observatory Virtual Museum
Timeline
People
Instruments
40-inch Refractor
Kenwood Observatory
12-inch Kenwood Refractor
24-inch Ritchey

The Largest Refracting Telescope in the World
By Patricia G. Roussel

Warner and Swasey received the order for the Yerkes 40-inch telescope in October of 1892, and by the following spring the mounting was on exhibit at the Columbian Exposition in Chicago. They were given the contract because of their excellence in workmanship, specifically because of the valuable experience gained by the firm in designing and constructing the mounting of the 36-inch refractor at Lick Observatory, a smaller

Figure 11.4. Patricia Roussel's history of the Yerkes 40 inch Refractor (see CD-ROM).

Mount Wilson Observatory (MWO), located just outside Pasadena, California, is home to the Hooker 100 inch telescope. When completed in 1917 this was the world's largest and remained so for the next 30 years until the Palomar 200 inch was completed.

Included

on CD-ROM

The web site has a virtual tour and information on how to request time on the 100 inch telescope. This document, which also gives much background information on the telescope, is included on the CD-ROM, as are details of public opening times. One novel service which the site offers is personalised Sky Maps. By filling in a form with your relevant particulars, your own Sky Map will be created (Figure 11.6, *overleaf*). The resultant map is however in postscript format for printing but some graphics packages will open it for viewing.

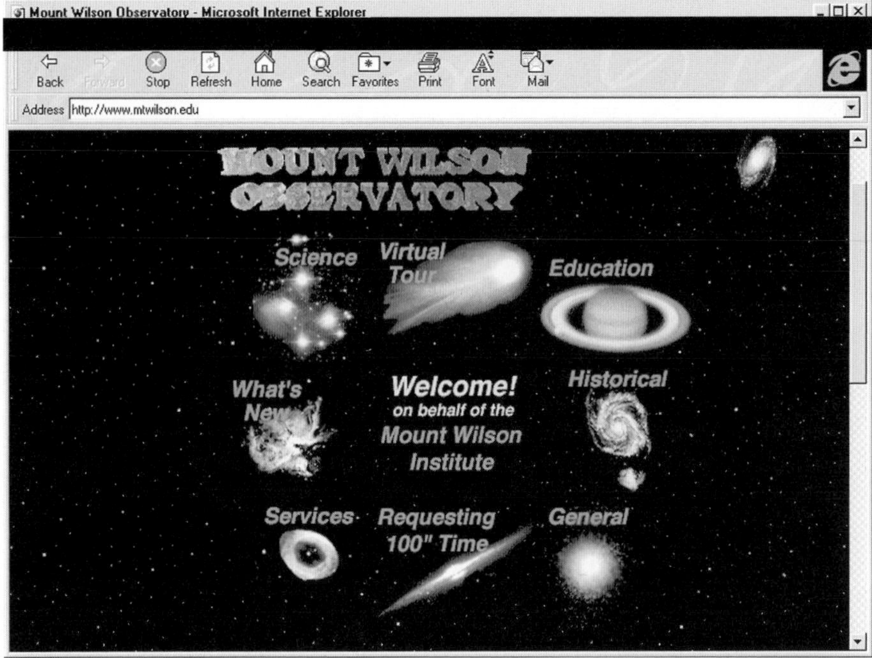

Figure 11.5. The Mount Wilson Homepage – home to the Hooker 100 inch Telescope.

Anglo-Australian Observatory

We now look at a more modern facility, the Anglo-Australian Observatory at Siding Spring, Australia. For more than 20 years this Observatory has operated two premier telescopes, the 3.9 metre Anglo-Australian Telescope and the 1.2 metre UK Schmidt Telescope. The latter was responsible for the first deep all sky survey of the southern skies. Located at Siding Spring it takes advantage of some extremely dark skies.

Much important research has taken place here on both telescopes but, as far as the public is concerned, it is most famous for the stunning astrophotographs of David Malin (Figure11.8, *overleaf*). Despite the advent of exotic solid state CCD cameras, David's film based photographs are still some of the most beautiful available and are much reproduced in magazines and books. But to appreciate them at their best, they need to be seen at poster size.

▶ **Figure 11.7.** The Anglo-Australian Observatory Homepage and home of the images of David Malin.

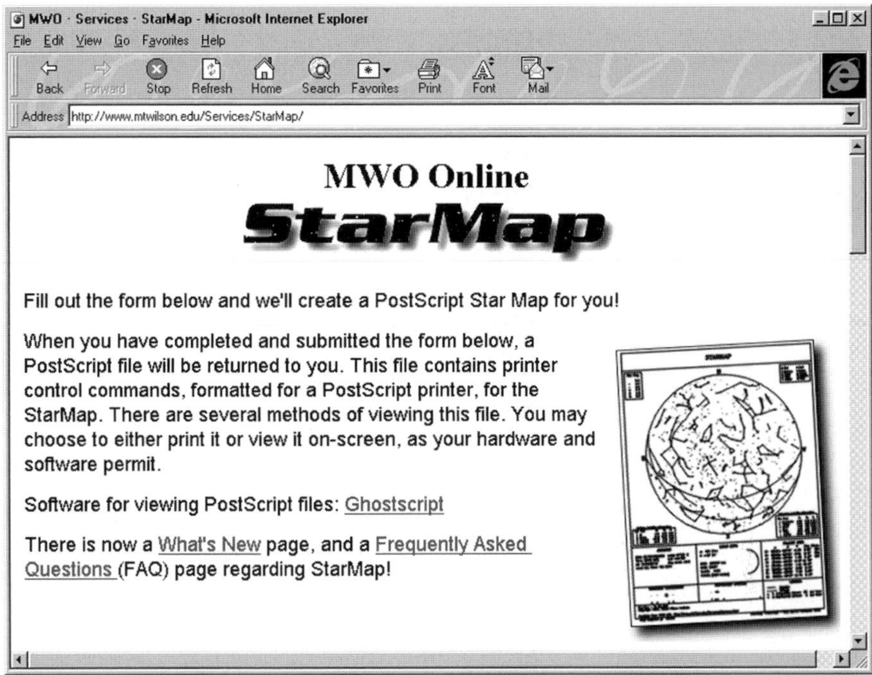

Figure 11.6. Personalised Star Maps at the Mount Wilson Observatory.

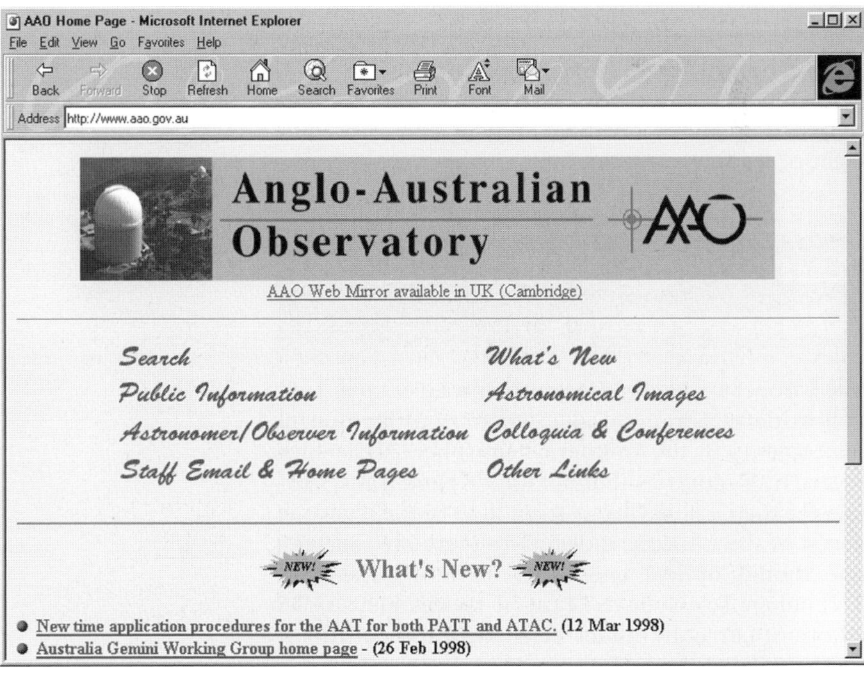

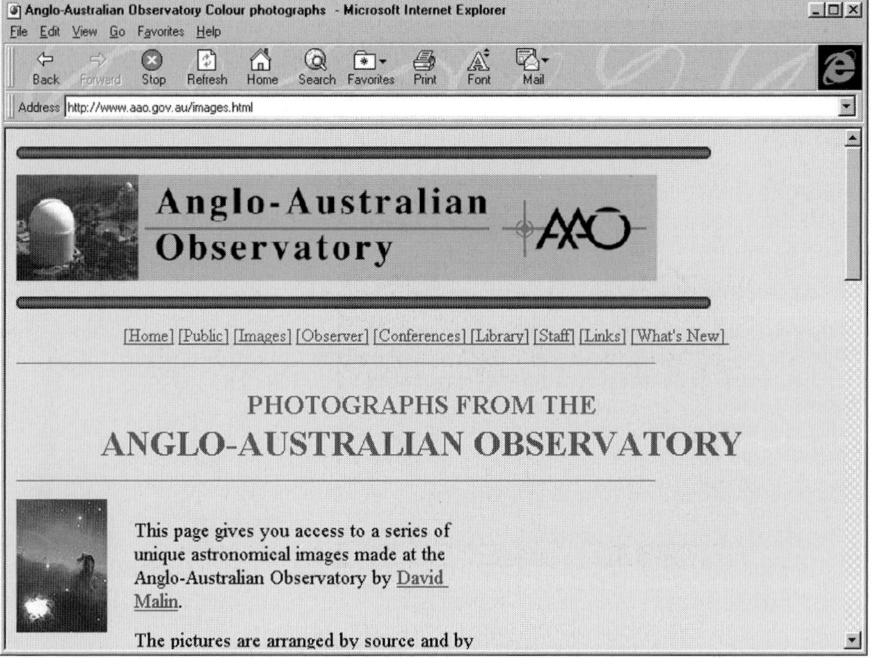

Figure 11.8. Gateway to the David Malin masterpieces!

Many of David Malin's spectacular images are available in low resolution for downloading. Details are also given of how to order full resolution copies.

An introduction to the observatory and its telescopes is included on the CD-ROM and is recommended reading.

European Southern Observatory

The European Southern Observatory was set up in 1962 to provide an astronomical observatory with powerful instruments in the southern hemisphere. It is supported by 8 countries (but not the UK) and is responsible for the La Silla Observatory high in the Atacama desert of the Chilean Andes. Here there are fourteen operational optical telescopes including the New Technology Telescope (NTT) at 3.5 metres aperture. In addition, currently under construction in northern Chile at Cerro Paranal, is the Very Large Telescope

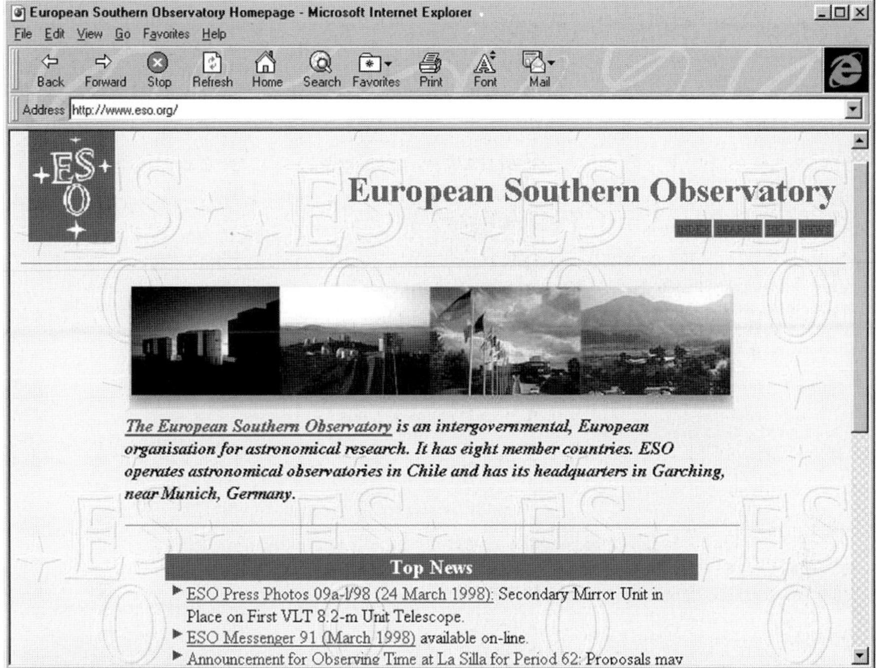

European Southern Observatory Homepage - Microsoft Internet Explorer
File Edit View Go Favorites Help
Back Forward Stop Refresh Home Search Favorites Print Font Mail
Address http://www.eso.org/
European Southern Observatory
The European Southern Observatory is an intergovernmental, European organisation for astronomical research. It has eight member countries. ESO operates astronomical observatories in Chile and has its headquarters in Garching, near Munich, Germany.
Top News
ESO Press Photos 09a-l/98 (24 March 1998): Secondary Mirror Unit in Place on First VLT 8.2-m Unit Telescope.
ESO Messenger 91 (March 1998) available on-line.

Figure 11.9. The European Southern Observatory (ESO) Homepage.

(VLT). The VLT will comprise an array of no less than four 8.2 metre telescopes which can operate individually or be combined to act as one giant one. In this latter mode it will be equivalent to a 16 metre telescope, making it what will be the largest optical telescope in the world when completed (Figure 11.10, *overleaf*).

Included on CD-ROM

Details of this exciting project are included on the CD-ROM. Two technical drawings are included in the information pack.

Jodrell Bank, The Nuffield Radio Astronomy Laboratories

In 1957 the Mark 1 radio telescope, at 76 metres in diameter, became the largest telescope of any kind in the world. Masterminded by Sir Bernard Lovell it became internationally famous for its astronomical research and its ability to track the USSR's space

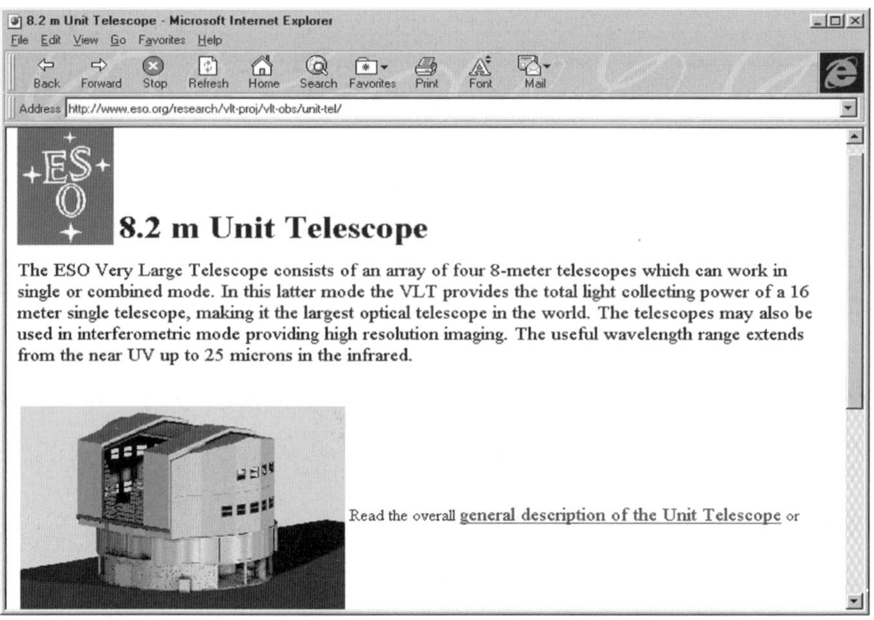

Figure 11.10. The ESO 8.2 metre telescopes currently under construction.

Figure 11.11. The Jodrell Bank Guided Tour.

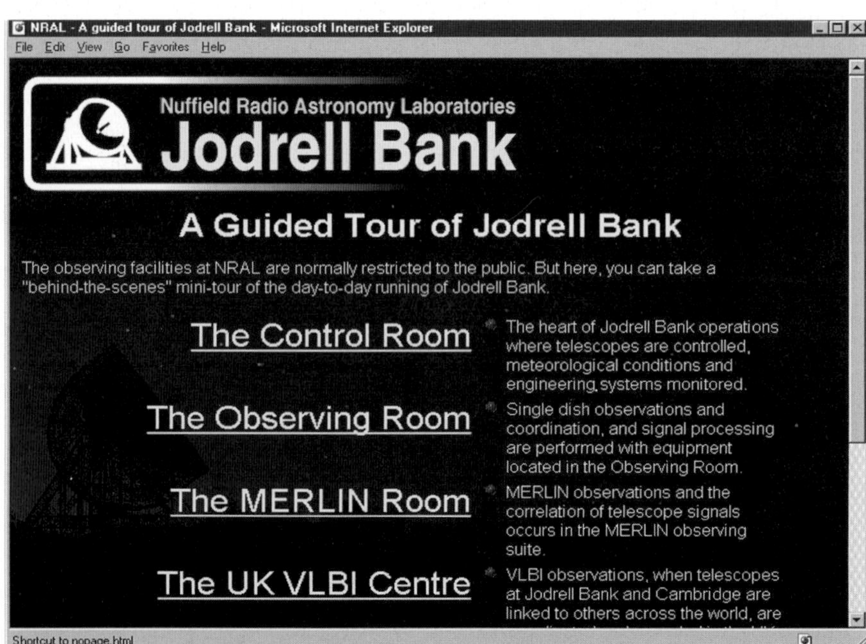

probes. Such was the interest in the site that in 1966 a visitor centre was opened. This has become not only the leading astronomical science centre in the North of England but a major tourist attraction as well.

However, it is the Lovell Radio Telescope (renamed from the Mark 1a) which grabs the attention. If you lived through the space race, then you will not need reminding of the high profile which Jodrell and Sir Bernard Lovell attained. Suffice it to say that they were both major celebrities, and were never out of the news. Since then, Jodrell has just got on with its work and with MERLIN (an array of radio telescopes all interferometrically linked) it is still at the leading edge of research today. The web site has recently been upgraded so that you can now take a virtual tour around the observatory and even "sit" at the control desk (Figure 11.12).

Included

on CD-ROM

Details of the history of the Lovell telescope, and Sir Bernard's role in its development, are included on the CD-ROM.

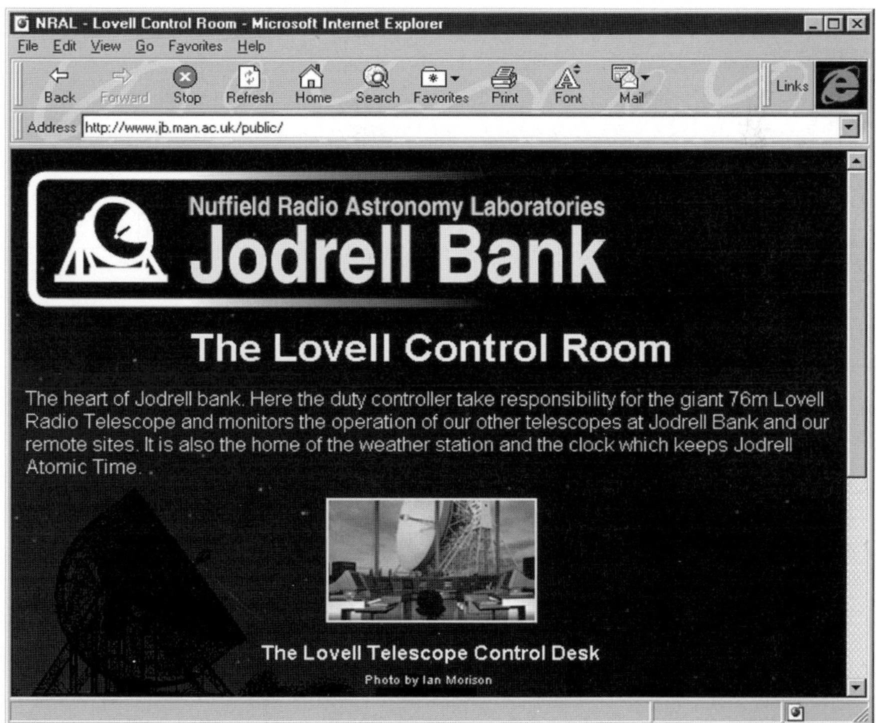

Figure 11.12. The famous Lovell Radio Telescope at Jodrell Bank as viewed from the Control Desk. Part of the Virtual Tour of the observatory.

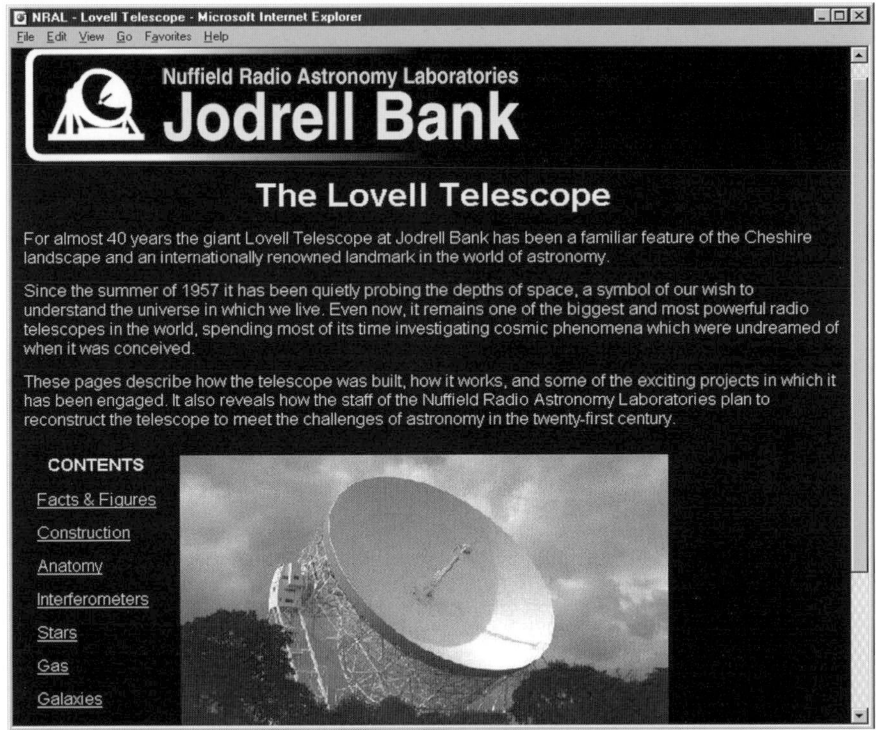

Figure 11.13. The Lovell Telescope – learn all about its history on the CD-ROM.

URLs Featured in this Chapter

Mount Palomar Observatory
http://www.astro.caltech.edu/observatories/palomar

Yerkes Observatory
http://astro.uchicago.edu/Yerkes.html

Mount Wilson Observatory
http://www.mtwilson.edu

Anglo-Australian Observatory
http://www.aao.gov.au

European Southern Observatory
http://www.eso.org

Jodrell Bank Radio Observatories
http://www.jb.man.ac.uk

Useful URLs not Referred to but Recommended

National Optical Astronomy Observatories
http://www.noao.edu

Kitt Peak National Observatory (including WIYN)
http://www.noao.edu/kpno/kpno.html

Gemini 8-Metre Telescope Project
http://www.gemini.edu

The Royal Observatory, Edinburgh
http://muinntiarach.roe.ac.uk

The Steward Observatory – Large Binocular Telescope
http://www.as.arizona.edu/lbtwww/lbt.html

The Bradford Robotic Telescope
http://www.telescope.org/rti

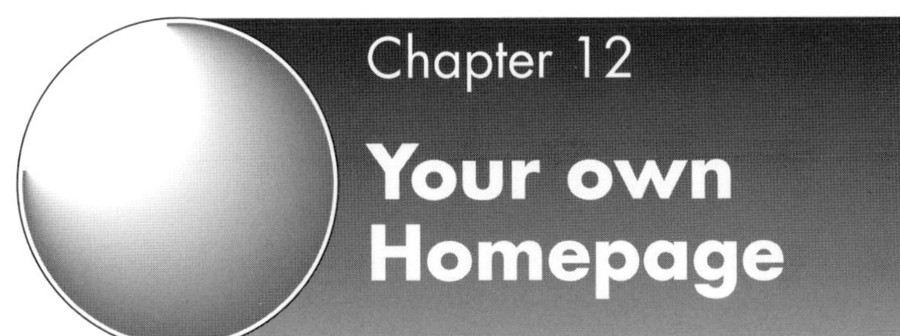

Chapter 12 Your own Homepage

Background

There is a publishing revolution going on. All over the world ordinary people, including amateur astronomers, are publishing their own material on the Internet. As amateur astronomers we have a hobby and interest that is worth sharing and what better way to do so than on your own web site.

Most Internet Service Providers (ISP), as part of their subscription package for their members, will give free web space for their own use. Even the most miserly ISP will offer 1Mb of space with the more generous offering 5Mb or even, in a few cases, 10Mb. Yet it is estimated that only 1 in 10 take up this offer. What puts them off is probably not knowing where start and how to go about it. There is plenty of jargon to put off the faint hearted, such at HTML and Java, but as we shall see these are nothing to be afraid of. So why not join the revolution and start publishing! Not only will you let the rest of the world know what you are doing but you could, as a result, get to know like minded astronomers all over the world.

Getting started

Most first generation web authors, as likely as not, started by plagiarising someone else's pages. Whilst plagiarism cannot (at least openly) be condoned it is a good way to learn. Web browsers have a "view source" option whereby the coding behind the web page can be viewed and printed. This source is in a coded language called HTML, Hypertext Mark Up Language. By viewing the source of a page which has taken your fancy, you can learn how that page has been created. At first it might appear off-putting but you should be able to recognise text and images which appear on the page. Rest assured everybody finds it strange at first but with a little perseverance HTML can be mastered. It is a bit like old fashioned word processing with embedded commands. These are surrounded by "pointy" brackets, and the structure is generally:

<command> text text text </command>

The first switches the command on, the second (note the /) switches it off. It is a good idea to either use upper or lower case consistently rather than mixing them up.

There are now software packages available which will generate HTML directly, such as FrontPage and Pagemill. However, although these might seem the easiest way to get going, in my opinion you are better starting with manual coding methods. There are several reasons for this but the primary one is control. Proprietary packages tend to take over and insist that you do it their way. If you then change the code to something different, the package could well change it back on its own!

To do-it-yourself the first step is to obtain a beginners guide to HTML. With this you can master the basics and produce quite a presentable homepage without worrying about fancy features at this stage (see Figure 12.1, *opposite*). The KISS principle (Keep it Simple, Stupid) applies to web pages as it does to most things in life. Included on the CD-ROM is just such a beginners guide from WebInfo and you are advised to print it out for easy reference. All you then need is a simple word processor to edit the HTML file. Remember to save it as an ASCII (text) with a name such as homepage.htm. This can then be viewed with your browser by asking it to open this file.

A common mistake with Homepages is to make the first page slow to load by including a lot of large graphics/images. The idea should be to encourage people to visit your pages and not to antagonise them with slow downloading. Better to have it loading quickly, welcoming visitors with a simple layout. This should get the message over logically and point to

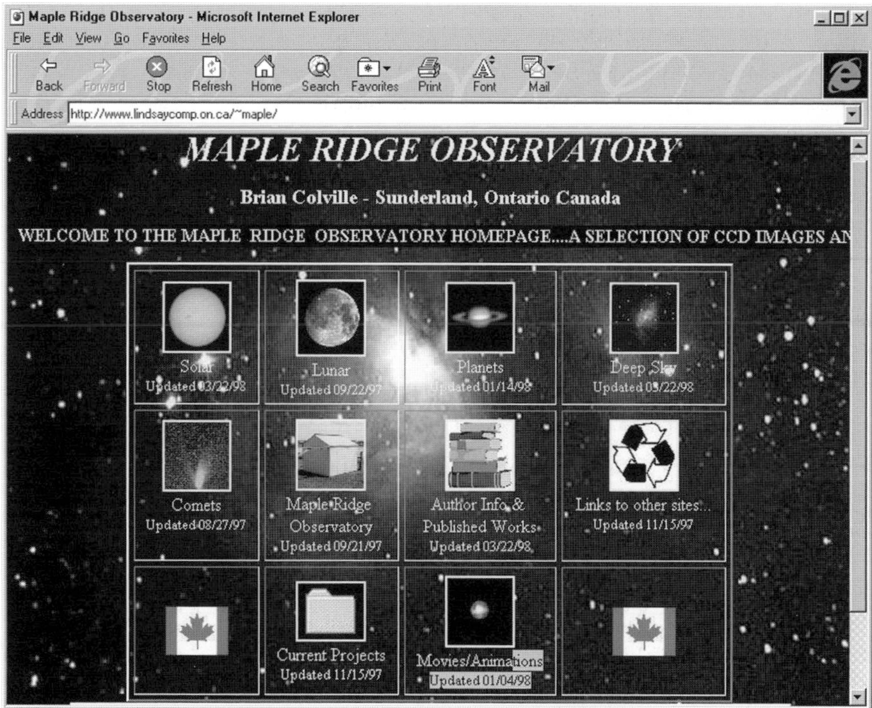

Figure 12.1. Brian Colville's Homepage – simple layout using small graphic images and a pleasing background.

where they can find out more about a topic of their choice. Get the visitor hooked first – then you can include your images and graphics.

Another point to remember is that visitors will have a mixture of screen resolutions. If you design the page on a 17 inch screen, you must check that it will view properly on a 14 inch one. Do this by shrinking the working window down and seeing how it looks. Not everyone has a high resolution monitor. I have had visitors to my web pages using Web TV – i.e. they are seeing them on a low resolution television.

Figure 12.2 (*overleaf*) is a very simple sample homepage which says who you are, what your homepage is about with easy to understand links to further topics. You are welcome to edit this with a word processor to suit your own purposes. Just put your name in and change the links to whatever topics you wish.

The HTML behind this page is as follows:–

```
<html>
<head>
<title>Joe Bloggs' Astronomy Homepage</title>
<meta http-equiv="Keywords" content="CCD Astronomy, CCD
Imaging, Telescope Making, Astrophotography">
</head>
```

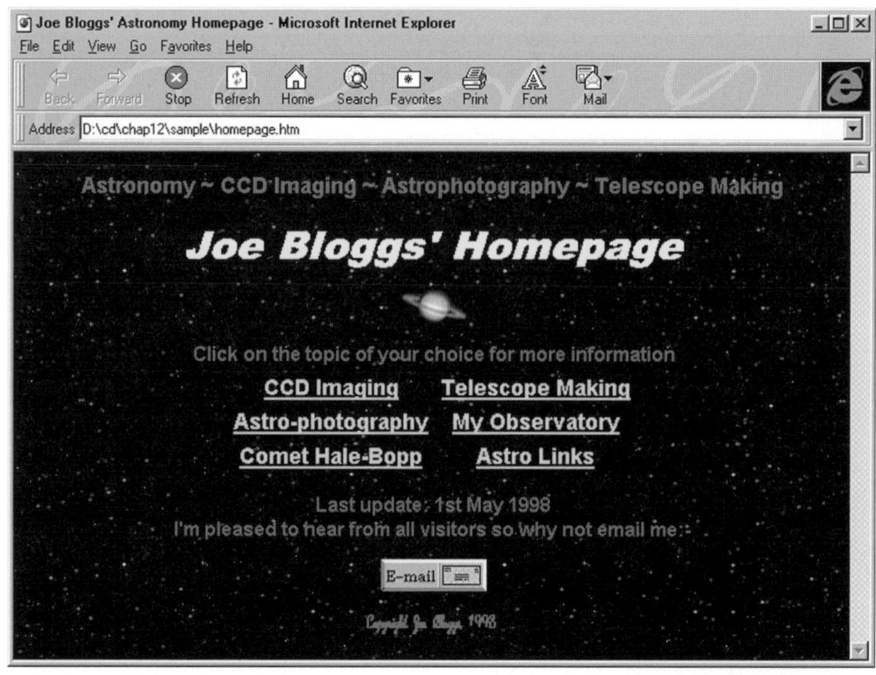

Figure 12.2. A simple homepage with clear links to further pages.

```
<body>
<body background="stars.jpg" bgcolor=black
text=red link=aqua alink=red vlink=blue>

<center><strong><font size=4 color="red" face="arial">Astronomy ~
CCD Imaging ~ Astrophotography ~ Telescope
Making</strong></font>
</center>
<P>

<center><strong><i><font size=6 color="yellow" face="arial
black">Joe Bloggs' Homepage</i></strong></font>
</center>
<P>

<center>
<img src="saturn.gif"></center>
<p>
<center>
<b><font size=3 color="red" face="arial">
Click on the topic of your choice for more information</b></center>

<center>
<table border=0 cellspacing=5>
<tr align=center>
<td align=center width=40%><a href="ccd.htm"><font
face="arial"size=4><b>CCD Imaging</b></font></a>
</td>
<td align=center width=40%> <a href="tm.htm"><font face="arial"
size=4><b>Telescope Making</b></font></a>
</td>
</tr>
```

```
<tr>
<td align=center width=40%><a href="photo.htm">
<font face="arial" size=4><b>Astro-photography</b></font></a>
</td>

<td align=center width=40%> <a href="obs.htm"><font face="arial"
size=4><b>My Observatory</b></font></a>
</td>
</tr>

<tr>
<td align=center width=40%>
<a href="comet.htm">
<font face="arial" size=4><b>Comet Hale-Bopp</b></font></a>
</td>
<td align=center width=40%> <a href="links.htm"><font face="arial"
size=4><b>Astro Links</b></font></a>
</td>
</tr>

</table>
</center>

<P>
<center><b><font size=3 color="red" face="arial">
Last update: 1st February 1998<br>

I'm pleased to hear from all visitors so why not email me:-</font></b>
<p>
<a href="mailto:you@yourhome ">
</center>

<center><img border=0 SRC="email.gif"></a><p>
<font size=3 face="script"><b>Copyright Joe Bloggs 1998</b></font>
</center>

</body>
</html>
```

To edit the file, open it with your word processor. Points to note are the title and keywords at the top. No one is quite sure how web search engines index the Internet (they keep their methods secret) but it could well be that title and keywords feature strongly, so it pays to include in them something highly relevant. After all you want surfers to find your pages. Edit "Joe Bloggs" to your name. The hypertext links to other pages will need changing if you have different interests to mine. The image included on the page is saturn.gif, change this to one of your own choice. Finally change the email address to your own. It is as simple as that. A background image would improve the overall look but remember to keep it very small if you decide to use one.

For the other pages to which the homepage links point, these can be similar but I would delete the background colour (black on the homepage) so that it defaults to white or grey. This is because text colours on black are not the best to read from and they do not print out well, in fact some printouts can be impossible to read. A point to bear in mind.

On subsequent pages you will no doubt want to include images of perhaps your telescope or observatory. These currently have to be in either GIF or JPG format. An alternative PNG is coming. I would recommend JPG for photographs as it can be highly compressed to make small quick loading images. However, every time you save a JPG image it degrades. So get everything right, size, contrast, cropping etc., before converting it finally to JPG format. The GIF format is better for graphs and drawings. Both these formats can be created with Paintshop Pro (Chapter 8).

Having created our web pages we can thoroughly test them on our hard disc. When you are satisfied that all the links work and they look good on a variety of screen sizes, then you are ready to upload them to your ISP's server. Included with your membership pack should be instructions and some utilities to do this. The files are uploaded in batch i.e. we highlight them all (HTML and graphic files) on our hard drive and transfer them into our directory on the ISP's server. You will need to be given a password by your ISP for this. Your ISP should provide compatible software to do all this.

Having uploaded our pages, there is one final task to do. Whilst web search engines will eventually find your homepage it is a good idea to give them a little help. Most of these have in the small print at the bottom of their homepage, an option for "submitting" URLs to them directly. My advice would be take advantage of this and submit your URL to several of the top search engines (see Chapter 1). Even then do not expect to be able to do a search the following day and have the search engines find you. It takes time unfortunately. Another route is to let your friends know you are online and offer them, in return for them providing a link to you, a link to their pages.

This has been a very brief introduction to HTML but once you have mastered the basics it is easy to start including more advanced features in your pages. Perhaps you could next include moving images or text. The HTML guide gives instructions on how to achieve these and you could soon find plenty to do on those cloudy nights. If a background image appeals then take a look below at Universe 1.5 and make your very own space scene, but remember to keep the file size small, otherwise visitors will not wait for it to load. I would suggest a file size no bigger than 10 k.

My own web pages are included on the CD-ROM as an example of a typical amateur astronomer's site. Most of the images have been heavily compressed to keep them small.

Diard's Universe Software

An eye-catching background can do wonders for a web page. Universe, written by Jess Diard, produces just such backgrounds and with an astronomical theme. It uses fractal techniques to generate these and as a result no two are identical. In the trial version we are limited to star-fields with clusters and nebula effects superimposed on top (Figure 12.4, *overleaf*). Each effect can be varied in a number of ways to personalise the picture. In addition you can paste in your own images, say a picture of Saturn. There is an option to bring these in with a transparent background, so they merge perfectly into the star-field. The full registered version adds planets and moons of its own for even more impressive space scenes plus "vortices" for black holes. The image type that it can

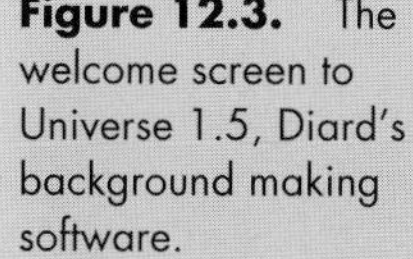

Figure 12.3. The welcome screen to Universe 1.5, Diard's background making software.

Universe
Version 1.5
Copyright © 1996-1997 Diard Software.
www.diardsoftware.com

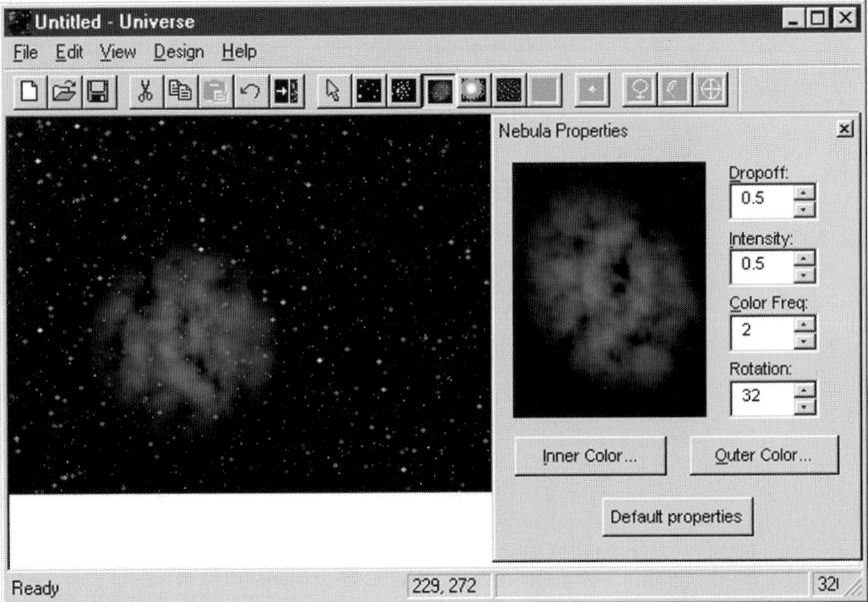

Figure 12.4. Adding a "nebula" to a star-field background using Universe 1.5.

import and save to, is limited to bitmap format (BMP), so another graphics package is required to convert them to a web compliant format, such as JPG or GIF.

The trial version is included on the accompanying CD-ROM and should you decide to register it, the locked features will become operable. There is an uninstall option with the software but I think you will want to keep it!

URLs Featured in this Chapter

Brian Colville's Maple Ridge Observatory Homepage
http://www.lindsaycomp.on.ca/~maple

Diard's Universe Software Homepage
http://www.diardsoftware.com

WebInfo Homepage – HTML Guide
http://www.webinfo.inch.com

Other Amateur Astronomer Homepages Worth Visiting:

Adrian Catterall
http://www.observatory.demon.co.uk

Tim Puckett
http://www.cometwatch.com

David Strange
http://ourworld.compuserve.com/homepages/dstrange

Nik Szymanek and Ian King
http://ourworld.compuserve.com/homepages/Nik_Szymanek

Gregory Terrance
http://www.frontiernet.net/~gregoryt

Luc Vanhoeck
http://ourworld.compuserve.com/homepages/Luc_Vanhoeck

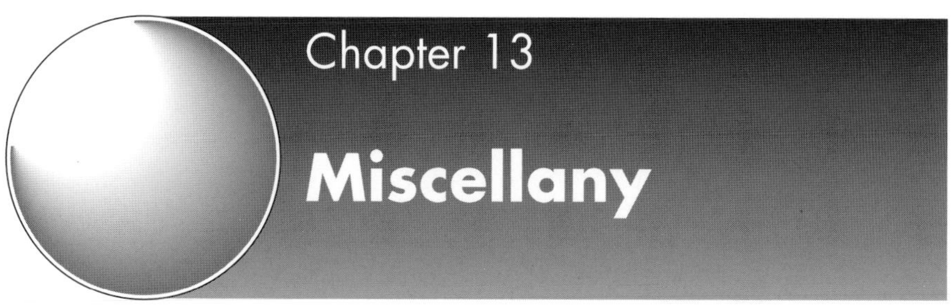

Chapter 13
Miscellany

Background

In compiling this book, I came across certain packages or web sites which were too good to leave out but did not easily fit in with the structure of the book. In addition, others were omitted from chapters simply because there was insufficient room. Hence this chapter where I have brought them together. Hopefully there should be something for everybody here.

H-R Calc

This shareware program for Windows (3.1 and up) was written by David C. Irizarry and is a little gem. Every amateur astronomer will have heard of the Hertzprung-Russell Diagram, but how many understand the principles behind this fundamental aspect of stellar mechanics? It demonstrates graphically the relationship between a star's theoretical radius, its luminosity and its spectral class (surface temperature). Built into it is the data for most bright stars which can be selected either by their name or by Greek letter and constellation. This is probably the easiest way to learn how to use the program. Once selected, the stars properties are displayed in their correct position on the H-R diagram. Slider bars are available for changing parameters and investigating their effects. Other displays are scale diagrams

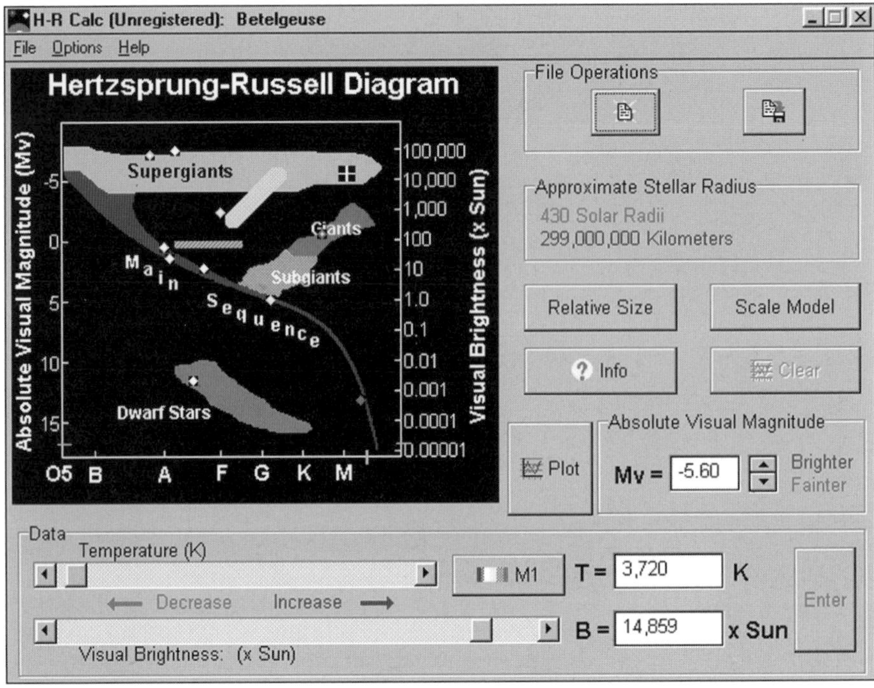

Figure 13.1. The H-R Calc display for the star Betelgeuse.

showing the stars size relative to the Sun and Earth's Orbit (Figure 13.2).

The shareware version is included on the CD-ROM and although not restricted to 30 days use as is the norm, each session is time-limited. You will have the bother of having to restart when your time runs out. Details of how to register and get rid of this minor irritation are included in the readme file.

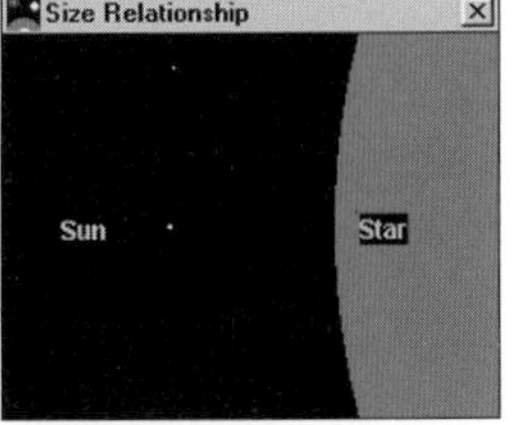

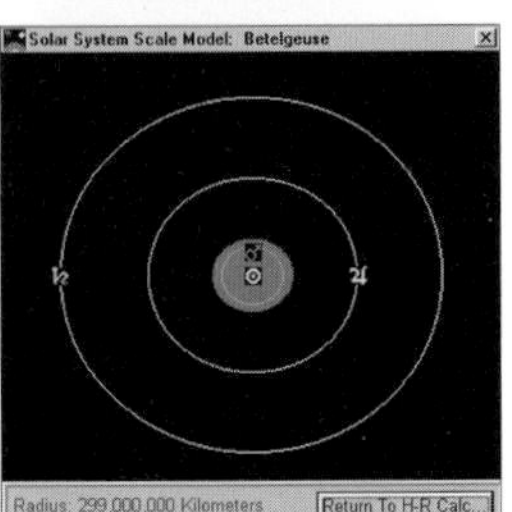

Figure 13.2. The relative size of Betelgeuse to the Sun and the Earth's orbit.

Astromart

I must admit that this is one of my regular sites to visit. These free classified advertisements are extremely popular amongst those with an eye for a bargain. There are tens of new adverts appearing every day. Items range from a single eyepiece to the large commercial telescopes, all at bargain prices. Whilst it is true that most of the advertisements are US based, many UK amateur astronomers are now advertising their surplus equipment here. In any case the fact that the items for

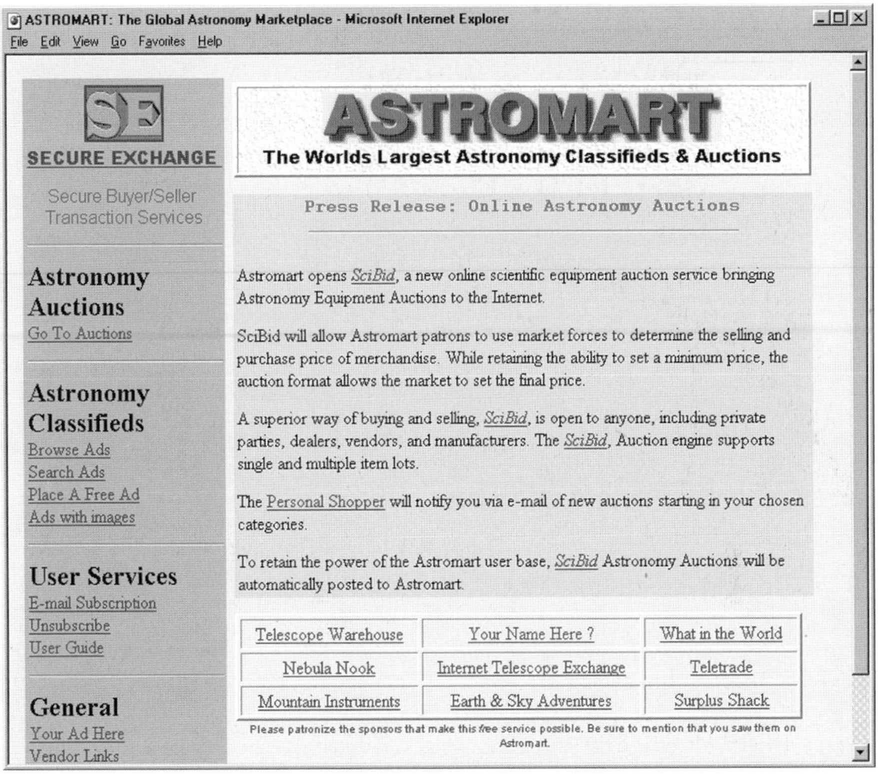

ASTROMART: The Global Astronomy Marketplace - Microsoft Internet Explorer
File Edit View Go Favorites Help

SE
SECURE EXCHANGE
Secure Buyer/Seller Transaction Services

Astronomy Auctions
Go To Auctions

Astronomy Classifieds
Browse Ads
Search Ads
Place A Free Ad
Ads with images

User Services
E-mail Subscription
Unsubscribe
User Guide

General
Your Ad Here
Vendor Links

ASTROMART
The Worlds Largest Astronomy Classifieds & Auctions

Press Release: Online Astronomy Auctions

Astromart opens *SciBid*, a new online scientific equipment auction service bringing Astronomy Equipment Auctions to the Internet.

SciBid will allow Astromart patrons to use market forces to determine the selling and purchase price of merchandise. While retaining the ability to set a minimum price, the auction format allows the market to set the final price.

A superior way of buying and selling, *SciBid*, is open to anyone, including private parties, dealers, vendors, and manufacturers. The *SciBid*, Auction engine supports single and multiple item lots.

The Personal Shopper will notify you via e-mail of new auctions starting in your chosen categories.

To retain the power of the Astromart user base, *SciBid* Astronomy Auctions will be automatically posted to Astromart.

Telescope Warehouse	Your Name Here ?	What in the World
Nebula Nook	Internet Telescope Exchange	Teletrade
Mountain Instruments	Earth & Sky Adventures	Surplus Shack

Please patronize the sponsors that make this *free* service possible. Be sure to mention that you saw them on Astromart.

Figure 13.3. The Astromart Homepage.

sale may be in the USA is no problem these days. I have personally bought two items through Astromart from the States without any problem. Prices tend to be lower there, even with postage and VAT (on import) you could well save a considerable sum. The items I purchased were a 2 inch diameter illuminated reticule eyepiece for \$45 and a Byers worm gear for \$100 which retailed at \$400! They took less than two weeks to arrive.

So popular has this site become that monthly advertisements have had to be split into 4 sections. Adverts can be sorted by date (Figure 13.4, *overleaf*) or by advertiser. There is also a search facility – useful if you are after a particular item.

There is a certain procedure and abbreviation jargon to learn, such as:

FS – for sale
WTB – want to buy
LNIB – like new in box

Included on CD-ROM

The guide to Astromart is included on the CD-ROM. If you have a weakness for bargains read it and start saving today!

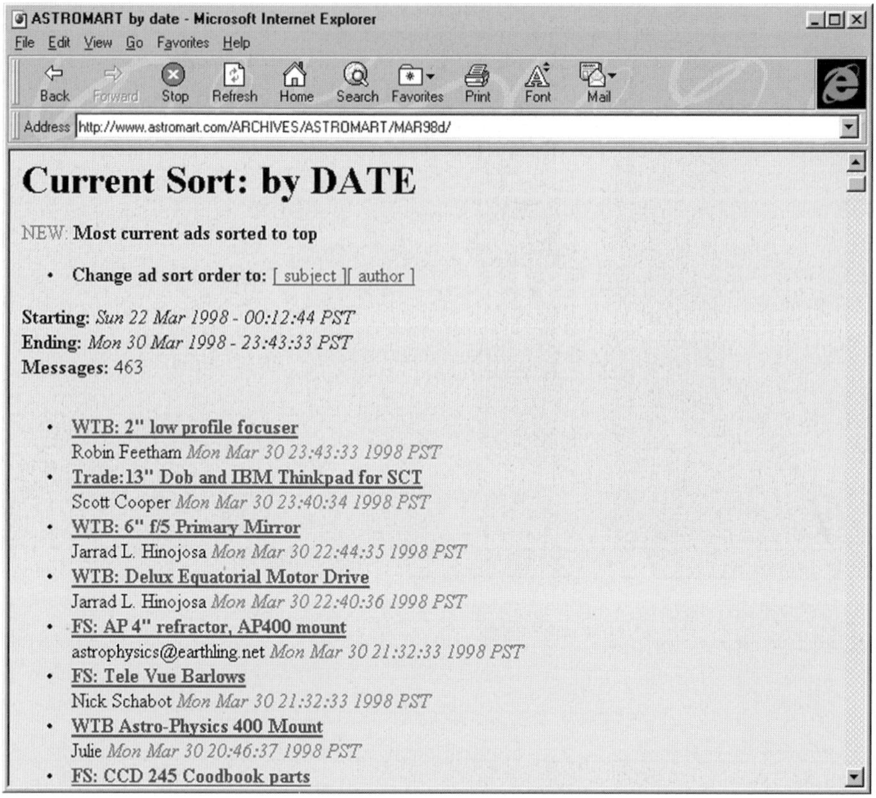

Figure 13.4. Astromart adverts sorted into date order.

Windows Dictionary of Astronomy

Included on CD-ROM

This software package, by the British company Pulsar, is a modernisation of an old idea, namely a dictionary of astronomy. The demo version (for Windows 3.1 or higher) included on the CD-ROM, is limited in extent, but its contents are nonetheless a useful source of background information. Whereas the full version has 2000 entries this one is limited to just over 100.

There are a variety of ways of searching the database such as by subject or free text search on any word. Also available are "Tables" and "Read Files" although they are again limited to a couple of entries in this version. "Tables" includes details of the 50 brightest star and basic data on the planets, The "Read Files" also include a variety of topics, the file for the eclipses is shown in Figure 13.6 (*opposite*). Additional reading material is

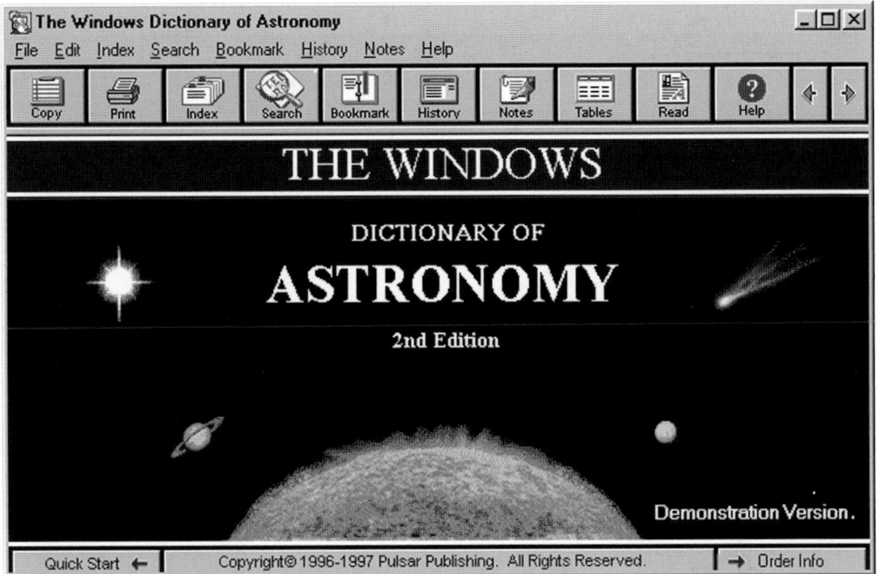

Figure 13.5. The opening screen for Pulsar's Dictionary of Astronomy.

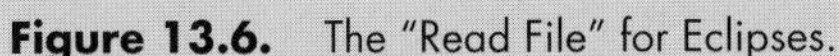

Figure 13.6. The "Read File" for Eclipses.

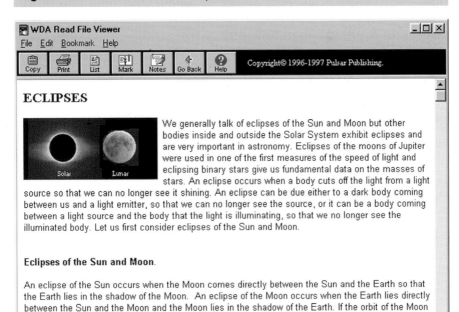

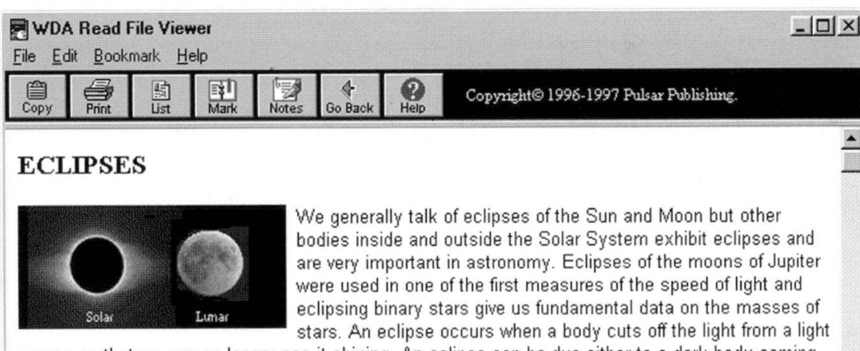

WDA Read File Viewer
File Edit Bookmark Help
Copy Print List Mark Notes Go Back Help
Copyright© 1996-1997 Pulsar Publishing.
ECLIPSES
Solar
Lunar
We generally talk of eclipses of the Sun and Moon but other bodies inside and outside the Solar System exhibit eclipses and are very important in astronomy. Eclipses of the moons of Jupiter were used in one of the first measures of the speed of light and eclipsing binary stars give us fundamental data on the masses of stars. An eclipse occurs when a body cuts off the light from a light source so that we can no longer see it shining. An eclipse can be due either to a dark body coming between us and a light emitter, so that we can no longer see the source, or it can be a body coming between a light source and the body that the light is illuminating, so that we no longer see the illuminated body. Let us first consider eclipses of the Sun and Moon.
Eclipses of the Sun and Moon.
An eclipse of the Sun occurs when the Moon comes directly between the Sun and the Earth so that the Earth lies in the shadow of the Moon. An eclipse of the Moon occurs when the Earth lies directly between the Sun and the Moon and the Moon lies in the shadow of the Earth. If the orbit of the Moon about the Earth lay in the same plane as the orbit of the Earth about the Sun then there would be eclipses of the Sun and Moon at every New and Full Moon respectively. The orbits are inclined, however, and eclipses can only occur when the Moon is close to the nodes of its orbit (when it is near to the places where the orbital planes cross).

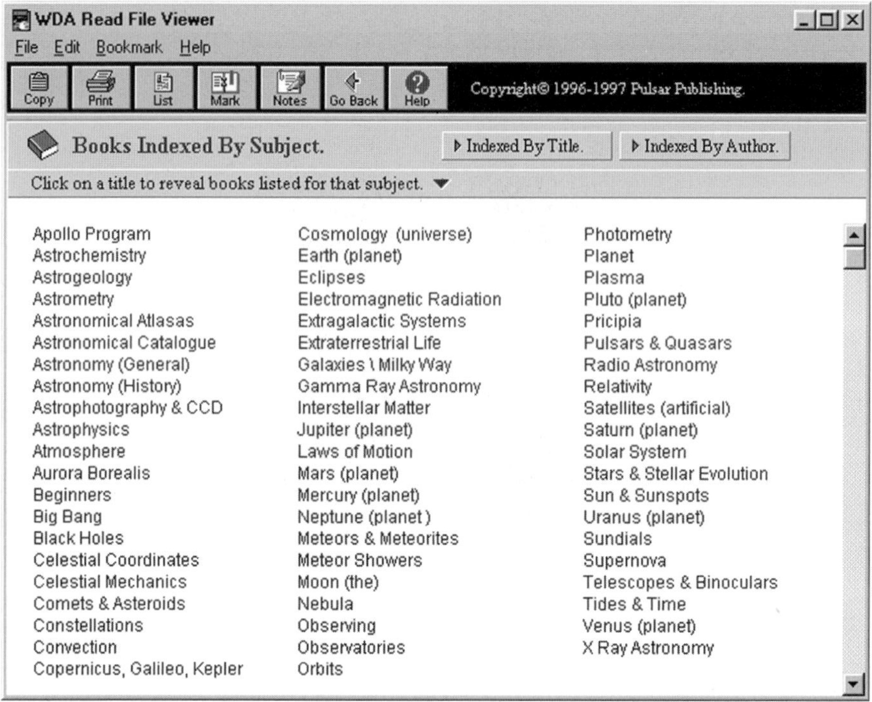

Figure 13.7. The book subject index.

also available in the form of a book list. This is searchable by subject only in the demo version but is quite comprehensive (Figure 13.7). Details of how to order the full version are included in the help files.

To install the program just click on the self installing file wdademo.exe and enjoy. The cost of the full version is a very reasonable £14.95 (or $23.95 in the US).

Messier Marathon

Have you ever wondered how many of the Messier Objects are visible from your location on a particular night? Well this little Windows application written by Bev Ewen-Smith, who runs the popular astronomical holiday observatory in Portugal (known as COAA), solves that problem. No doubt he must have been asked on several occasions by visitors to his observatory, “Which “M” objects are visible tonight?” – so wrote this to answer that question.

Although not over endowed with help facilities it is relatively easy to muddle your way through the program. Firstly you need to enter your location. You

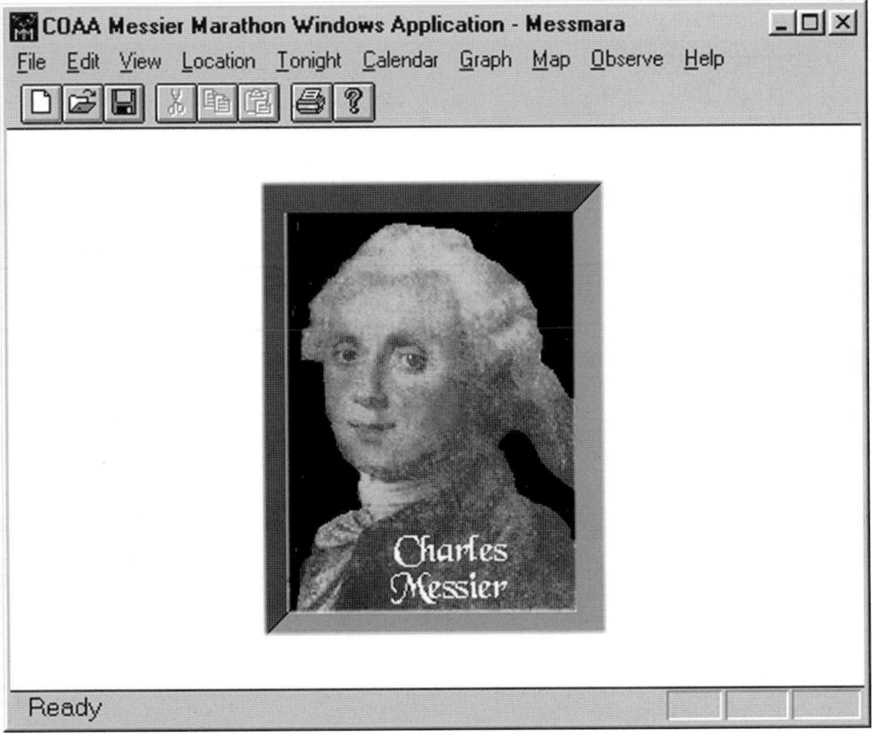

Figure 13.8. Bev Ewen-Smith's Messier Marathon program.

can then click on "Tonight" and the number of Messier objects visible will be displayed, together with a list of those which are NOT visible – obviously shorter than listing those which are (Figure 13.9).

Suppose you want to observe the maximum possible, which is the best night of the year? The program provides this information in tabular format and, probably more usefully, graphically as well (Figure 13.10, *overleaf*). For my location it turns out that the best I can hope for is 101 objects on one day in March. Using the

Figure 13.9. Visible tonight are 99 Messier Objects plus a list of those NOT visible.

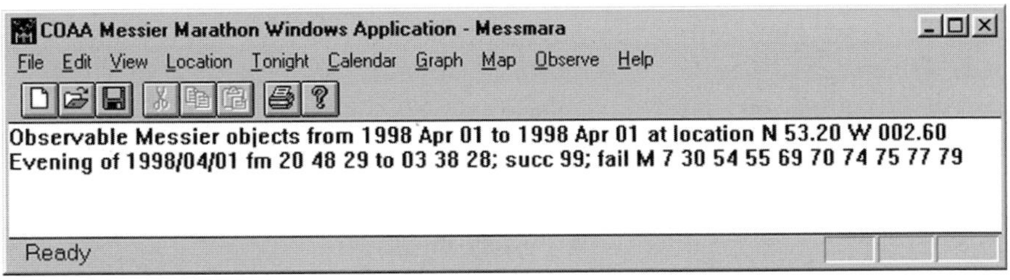

COAA Messier Marathon Windows Application - Messmara
File Edit View Location Tonight Calendar Graph Map Observe Help

Observable Messier objects from 1998 Apr 01 to 1998 Apr 01 at location N 53.20 W 002.60
Evening of 1998/04/01 fm 20 48 29 to 03 38 28; succ 99; fail M 7 30 54 55 69 70 74 75 77 79

Ready

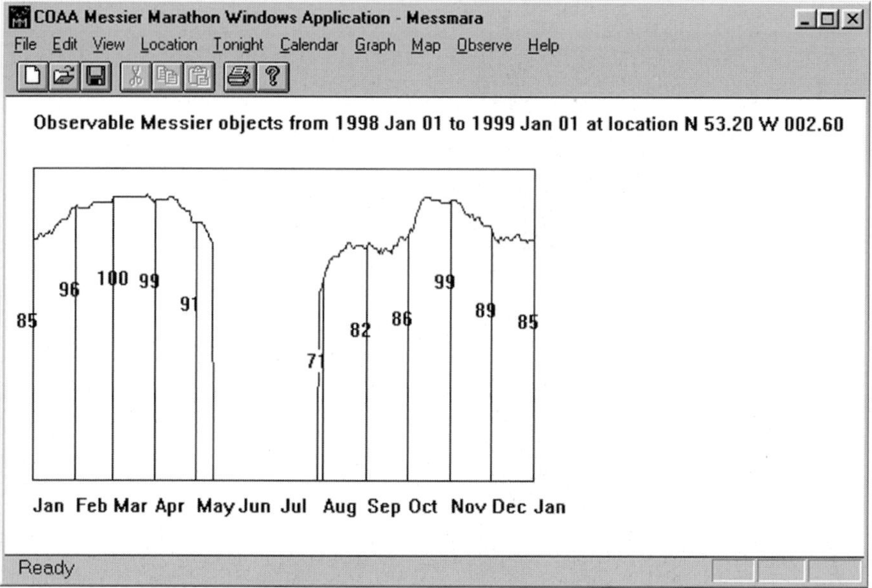

"Calendar" option, I can determine that the day in question is 26th March. I wonder if it will be clear?

To run this freeware program just click on Messmara.exe and remember you owe a vote of thanks to Bev Ewen-Smith.

Figure 13.10. Graph of the number of Messier Objects visible for a location in the North of England.

Visual Planets 2.0

Included

on CD-ROM

Visual Planets is really an educational program providing a tour of the Sun and its nine planets. It is however of interest to the amateur astronomer needing a useful reference to the solar system. Making extensive use of hypertext, it is easy to navigate through a combination of text and images to find out about a planet's moons, atmosphere, structure, rings (if present), missions, statistics and mythology. In this shareware version the information for Mars, Saturn, Neptune and Pluto is omitted but Figures 13.12 (*opposite*) and 13.13 (*overleaf*) show just some of the information it does provide. From the same author as Windows Dictionary of Astronomy it includes 100 definitions taken from that program.

It requires Windows 3.1 or higher to run and includes an un-installer should you decide not to keep it. The price for the full registered version is a reasonable £14.95 ($23.95) and details of how to register are included in a readme file. To install it click on Vsshar20.exe.

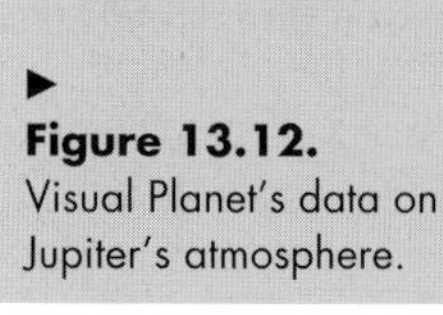
► **Figure 13.12.** Visual Planet's data on Jupiter's atmosphere.

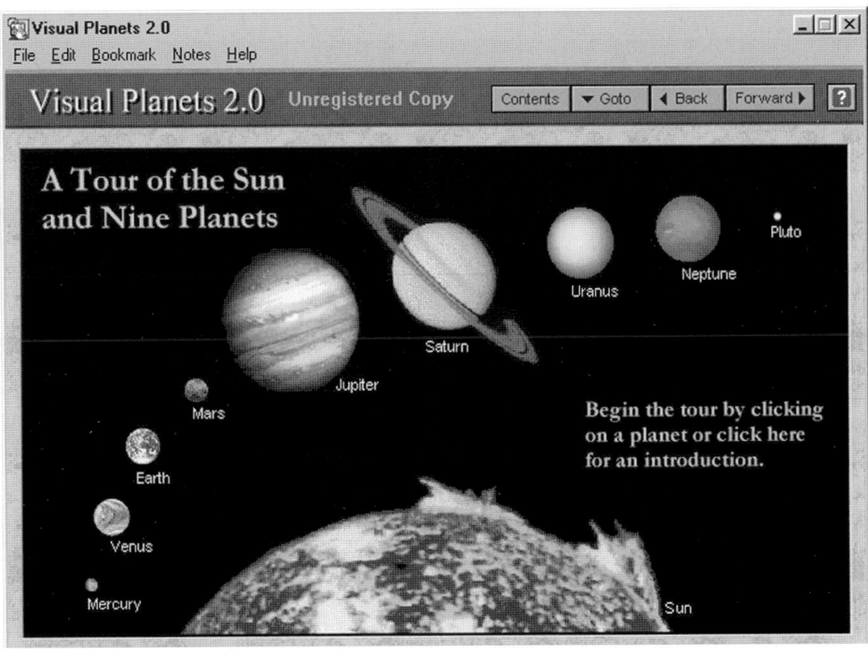

Figure 13.11. The opening screen from Visual Planets.

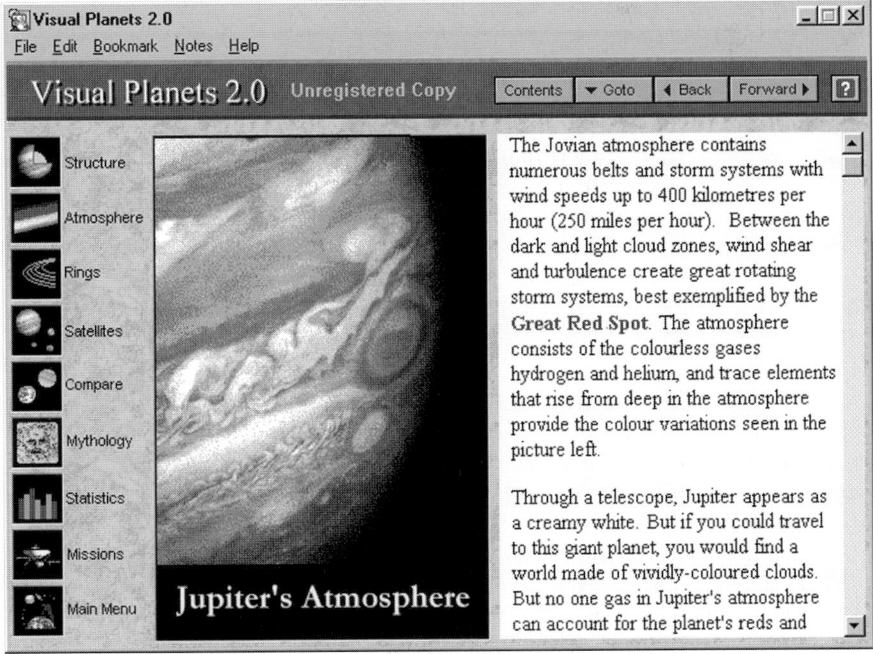

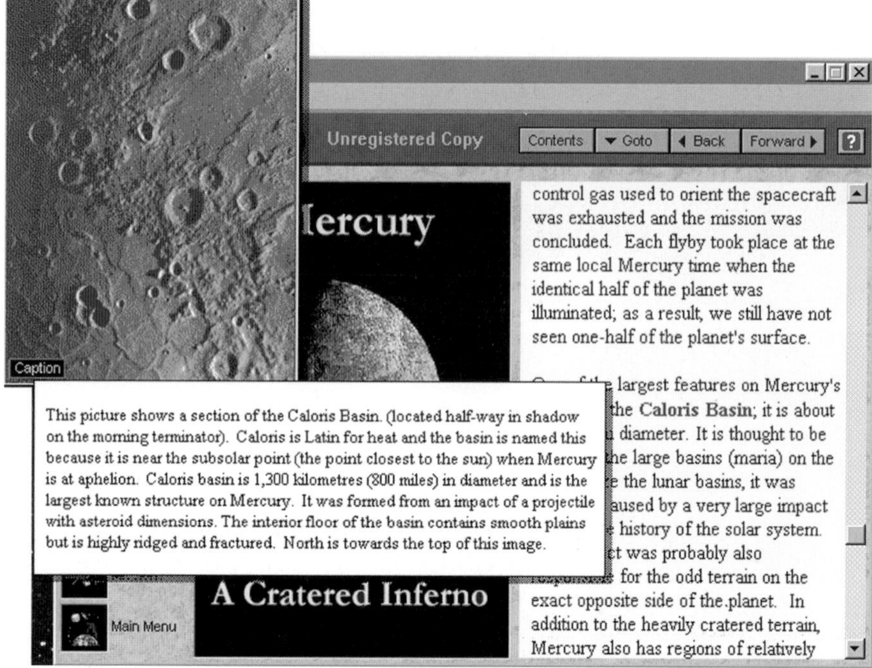

control gas used to orient the spacecraft was exhausted and the mission was concluded. Each flyby took place at the same local Mercury time when the identical half of the planet was illuminated; as a result, we still have not seen one-half of the planet's surface.

...largest features on Mercury's ...he Caloris Basin; it is about ...u diameter. It is thought to be ...he large basins (maria) on the ...e the lunar basins, it was ...aused by a very large impact ...history of the solar system. ...t was probably also ...for the odd terrain on the exact opposite side of the planet. In addition to the heavily cratered terrain, Mercury also has regions of relatively

This picture shows a section of the Caloris Basin. (located half-way in shadow on the morning terminator). Caloris is Latin for heat and the basin is named this because it is near the subsolar point (the point closest to the sun) when Mercury is at aphelion. Caloris basin is 1,300 kilometres (800 miles) in diameter and is the largest known structure on Mercury. It was formed from an impact of a projectile with asteroid dimensions. The interior floor of the basin contains smooth plains but is highly ridged and fractured. North is towards the top of this image.

A Cratered Inferno

Main Menu

Figure 13.13. Visual Planet's data for Mercury.

URLs Featured in this Chapter

Astromart – classified advertisements
http://www.astromart.com

Pulsar Publishing (Windows Dictionary of Astronomy & Visual Planets)
http://www.astrosoftware.com

The COAA Observatory software pages
http://www.algarvenet.pt/coaa/software.htm

Other URLs not Featured but Recommended

History of Astronomy
http://www.astro.uni-bonn.de/~pbrosche/astoria.html

Royal Astronomical Society
http://www.ras.org.uk/ras

Astronomy Picture of the Day
http://antwrp.gsfc.nasa.gov/apod/astropix.html

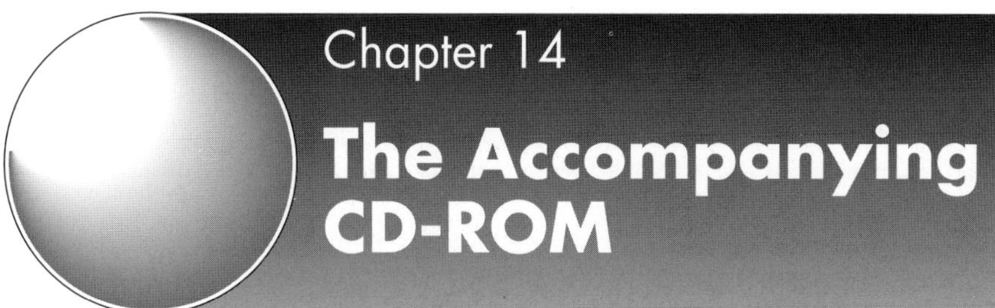

Chapter 14
The Accompanying CD-ROM

Background

Whilst it is relatively easy to ignore this chapter and dive straight into the CD-ROM, it will probably result in less frustration if you try it our way first. Firstly the good news, the CD-ROM is organised exactly as the book with sub-directories matching each chapter. If you read about something you fancy in Chapter 5 say, then it will be located in sub-directory \chap5. To help you get at the data and software a menu system is included. Now the bad news. For this menu system to run you must first install Adobe Acrobat Reader from the CD-ROM, unless you already have it installed on your PC. Once installed all you will need to do is double click on start.pdf located in the top directory of the CD-ROM (use explorer to do this). You can of course make a short-cut to simplify this and I would recommend doing so.

Installing Acrobat Reader

Two versions of Acrobat Reader are included on the CD-ROM. They are to be found in the Acrobat directory. Both are version 3.0 but one is a 16 bit version for Windows 3.1 and one a 32 bit version for Windows 95/98. Explore the appropriate sub-directory (\16bit or \32bit) for your operating system and click on the exe file. As with all Windows software you should not have any other applications running in the background. Acrobat Reader is a relatively small application and should only take up about 5MBs of disk space. You will need it for much of the NASA information anyway, so you might as well install it now.

As already mentioned, by clicking-on Start.pdf in the top level directory of the CD-ROM, the menu system should fire up. If not you may be prompted to associate an application with the pdf file type. If so, browse for Acroread.exe and that should do the trick. It sounds worse than it will be and if all else fails, startup Acrobat Reader and open the file Start.pdf on the CD.

Running the Menu System

You should now see something like Figure 14.1. If you are not familiar with Acrobat you should experiment with the icons and get the feel of what they do. Do not worry, you cannot break anything! However, as the start menu is a single page, the backwards and forwards

Figure 14.1. The CD-ROM menu in Acrobat Reader.

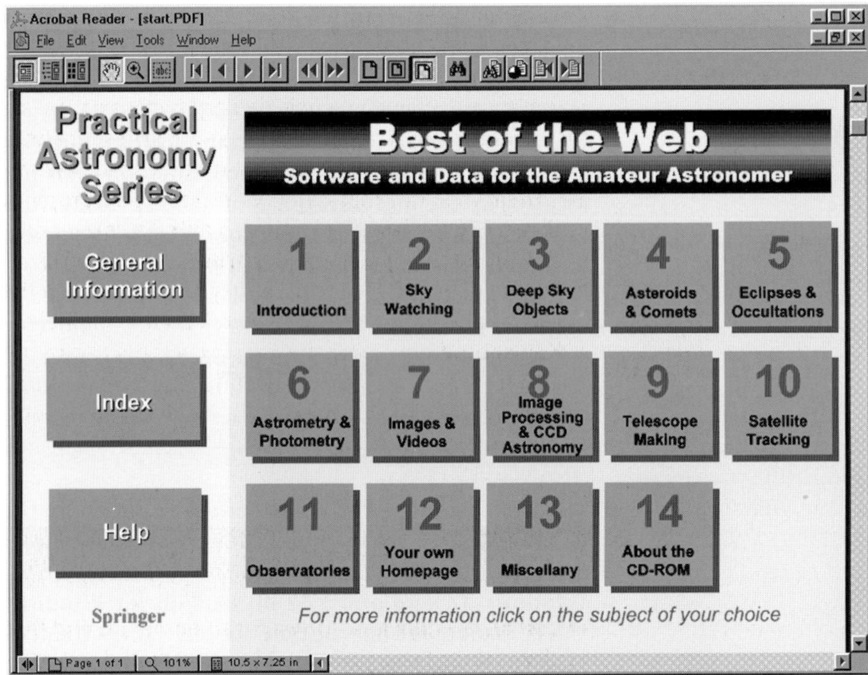

buttons will be inoperative – they come in when viewing multi-page documents. Points to note are that the mouse icon is normally a "flat hand" but when it passes over a hyperlink it changes to a "pointing finger". The binocular icon is for searching text in the document.

As you will see, the CD is organised into sections numbers 1 to 14 matching the chapters of the book. Acrobat is "point and click" so to choose a chapter just click on the box of your choice. This will enable you either to get to the data you want or, in the case of software, tell you how to go about running or installing it. Because of the risks involved in installing software with another application running (i.e. Acrobat), once you have read the instructions you will have to leave Acrobat for the actual installation. Data on the other hand can be accessed directly from Acrobat and there will be no need to quit. Many of the Hubble Space Telescope Press Releases are published in Acrobat format so having the menu system written in Acrobat makes sense. For those with very slow CD-ROM players (2× or even 4×) some of the bigger files will take a long time to load – just be patient. If they are still too slow, copy them to your hard drive.

Some of the information on the CD-ROM will require another application to run, such as a word processor, a web browser or MPEG movie player for videos. If you already have these applications then they should start up automatically and you will be asked to confirm that it is OK to do so – click on "yes". If they do not start up automatically then you will be invited to "open with" or "associate" the particular file type with an application. For images (GIF and JPG), I find it quicker to "associate" them with Internet Explorer rather than a graphics package. Use the browser option to find the correct one.

Included

on CD-ROM

If you do not have a web browser (e.g. Internet Explorer) or MPEG player then they are included on the CD-ROM and you will need to install them separately.

You should now have the necessary tools to run the software, view the data and watch the videos. I hope you enjoy them. I certainly did in compiling this book and CD-ROM.

Finally

Whilst all the data and software ran on my PC and that of the testers (and presumably the original author's too) there can be no certainty that it will run faultlessly on a particular PC. Such are the vagaries of Windows and system conflicts that no guarantee can be given. However, if you have any trouble then by all means you can email myself. I may not know the answer but I might know someone who does. Good luck!

My email address is:
david_ratledge@compuserve.com

and the URL of my web site is:
http://ourworld.compuserve.com/homepages/david_ratledge

For further information on Adobe Acrobat:
http://www.adobe.com

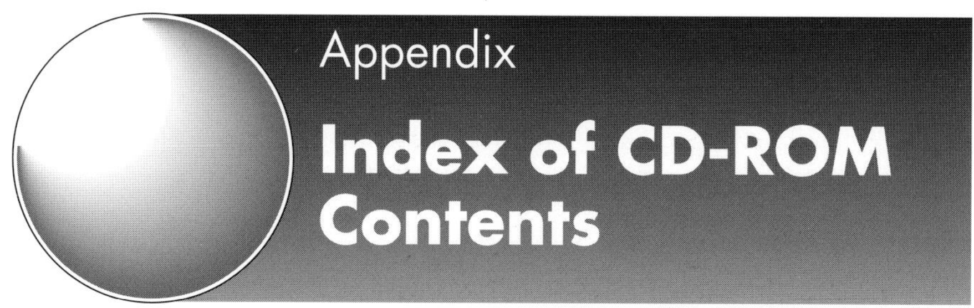

Appendix
Index of CD-ROM Contents

Adobe Acrobat Reader Software
Folder: \acrobat

\16bit	– Reader for Windows 3.1
\32bit	– Reader for Windows 95/98/NT

Chapter 1 – Introduction
Folder: \chap1

\Stuffit	– Stuffit Expander software
\Winzip31	– Winzip for Windows 3.1 software
\Winzip95	– Winzip for Windows 95/98 software
\gzip	– gzip software
\Starlink	– FAQ

Chapter 2 – Virtual Sky Watching
Folder: \chap2

\skymap22	– SkyMap2.2 Full Registered Copy
\skymap3	– SkyMap3.1 Shareware Version
\cybersky	– Cybersky software
\cosmohd	– Cosmohood software
\skytimes	– Skytimes software
\skyglobe	– Skyglobe software

Chapter 3 – Data Sources I: Deep Sky Objects
Folder: \chap3

\ngc	– NGC/IC Puzzle
\adc	– ADC FAQ
\sacds	– SAC Deep Space Database and software
\sacdbl	– SAC Double Stars Database and software
\sacfiles	– SAC Additional Databases
\dobjects	– dObjects software
\ads	– ADS Introduction

Chapter 4 – Data Sources II: Planets, Asteroids and Comets
Folder: \chap4

\horizons	– Horizons – Guide
\minplan	– Minor Planets (Asteroids) – Lists and Plots
\comets	– Comets – Lists and Plots
\mooncslc	– Mooncalc software
\cometwatch	– CometWatch, Images and Videos

Chapter 5 – Data Sources III: Eclipses and Occultations
Folder: \chap5

\eclipses	– Eclipses, Fred Espenak's Guides
\eclips99	– Eclipse 1999, Fred Espenak and Jay Anderson's Guide
\iota	– IOTA Introduction and Grazing Occultation Guide
\lowsta	– Lunar Occultation Workbench – standard edition
\lowcom	– Lunar Occultation Workbench – complete edition

Chapter 6 – Astrometry and Photometry
Folder: \chap6

\hipparc	– Hipparcos – Section 3.Tables plus other data
\astromet	– Astrometrica Software
\coaaorb	– COAA Orbit Software
\ezphot	– Ezphot – Software

Chapter 7 – Images & Videos
Folder: \chap7

\messier	– Digitised Sky Survey, sample Messier images.
\hstimage	– HST Public Images
\hstpdf	– HST Public Information – Acrobat format
\videos	– Videos
\galileo	– Galileo images
\marspath	– Mars Pathfinder – images
\marssurv	– Mars Global Surveyor – images
\earth	– Earth from Space – images

Chapter 8 – Image Processing and CCD Astronomy
Folder: \chap8

\cookbook	– Cookbook CCD Camera – FAQ
\fitsview	– Fits File Viewer software
\winmips	– WinMiPs Image Processing software
\quantum	– QuantumImage Image Restoration software
\psp	– Paintshop Pro Graphics Package software

Chapter 9 – Telescope Making
Folder: \chap9

\ukatm		– UK ATM Resources, Suppliers List and Email FAQ
\software		
	\cell	– Cell David Chandler's Flotation Cell Design
	\fouc_02	– Jim Burrow's Foucault Test Analysis
	\gsum	– Raytrace program for doublets
	\irt53	– Comprehensive raytrace program
	\mel	– Mel Bartels' Newtonian secondary mirror analysis

	\millac	– Spreadsheet for foucault testing
	\newt20	– Dale Keler's Newtonian CAD program (DOS)
	\newt25	– Dale Keler's Newtonian CAD program (WIN)
	\suiter	– Dick Suiter's suite of ATM software
	\tex23	– Foucault Test data reduction
	\vign9502	– Gerald Pearson's Newtonian Vignetting
\mel		– Mel Bartels – Ultra-light Dobsonian and motorised drives
\sayre		– Bruce Sayre's 20 inch Alt-Az Telescope
\bolton		– The Bolton Group's Telescopes – From 16 inch Newtonian to 8 inch binoculars
\tpoint		– Patrick Wallace – Tpoint and Telescope Pointing
\comsoft		– Comsoft – PC-TCS Telescope Control System demo software

Chapter 10 – Satellite Tracking
Folder: \chap10

\traksat	– Traksat software
\satspy	– Satspy software
\Satpro	– Satpro software

Chapter 11 – Observatories
Folder: \chap11

\palomar	– Mount Palomar and the 200 inch Telescope
\yerkes	– Yerkes Observatory and the 40 inch Refractor
\mtwilson	– Mount Wilson and the 100 inch Telescope
\aao	– Anglo-Australian Observatory – home of David Malin
\eso	– European Southern Observatory – Very Large Telescope project
\jodrell	– Jodrell Bank and the Lovell Telescope

Chapter 12 – Your Own Homepages
Folder: \chap12

\html	– Webinfo's HTML Guide
\sample	– Sample Pages
\drhome	– David Ratledge's web pages
\universe	– Diard's Universe Version 1.5 software

Chapter 13 – Miscellany
Folder: \chap13

\hrcalc	– H-R Calc, Hertzprung-Russell Diagram Software
\astromar	– Astromart Classified advertisements FAQ
\dictiony	– Windows Dictionary of Astronomy – software
\messmara	– Messier Marathon – software
\visplan	– Visual Planets – software

Chapter 14 – The Accompanying CD-ROM
Folder: \chap14

\iexplore	– Microsoft's Internet Explorer software
\mediaplayer	– MPEG movie players for Win95 and Win3.x

About the CD-ROM

The CD-ROM is organised exactly as the book with sub-directories matching each chapter. If you read about something in Chapter 5 say, then it will be located in sub-directory \Chap5. To help you get at the data and software a menu system is included. For this menu system to run you must first install Adobe Acrobat Reader Version 3 from the CD-ROM, unless you already have it installed on your PC. If you have an earlier version you are advised to upgrade to this one.

Installing Acrobat Reader

Two versions of Acrobat Reader are included on the CD-ROM. They are to be found in the Acrobat directory. Both are version 3.0 but one is a 16 bit version for Windows 3.1 and one a 32 bit version for Windows 95/98. Explore the appropriate sub-directory (\16bit or \32bit) for your operating system and click on the "exe" file. As with installing all Windows software you should not have any other applications running in the background. Acrobat Reader is a relatively small application and should only take up about 5MBs of disk space.

Then, by clicking-on Start.pdf in the top level directory of the CD-ROM, the menu system should fire up. If not, you may be prompted to "open with" an application for the pdf file type. If so, browse for Acroread.exe and that should do the trick. If all else fails, start Acrobat Reader and open the file Start.pdf on the CD-ROM.